D0687257

URBAN-RURAL CONFLICT

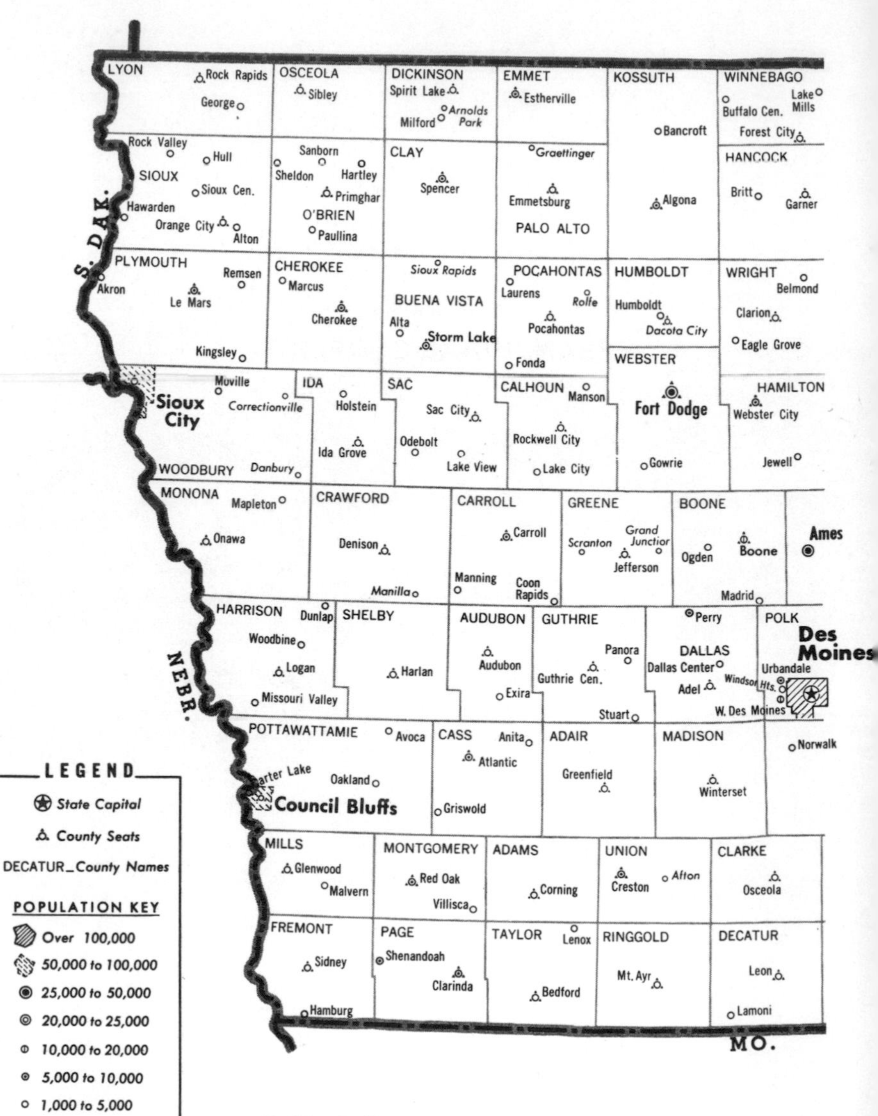

Population classification based on the Federal Census of 1960, corrected to latest state report.

COPYRIGHT ©
AMERICAN MAP COMPANY, INC.

Urban-Rural Conflict

THE POLITICS OF CHANGE

Harlan Hahn

University of California, Riverside

SAGE Publications
Beverly Hills, California

UNIVERSITY LIBRARY
Lethbridge, Alberta
90512

Copyright © 1971 by Sage Publications, Inc.

All rights reserved. No part of this book may be reproduced or utilized in any form or by any means, electronic or mechanical, including photocopying, recording, or by any information storage and retrieval system, without permission in writing from the Publisher.

For information address:

SAGE PUBLICATIONS, INC.
275 South Beverly Drive
Beverly Hills, California 90212

Standard Book Number: 8039-0080-5

Library of Congress Catalog Card Number: 73-127985

Printed in the United States of America

ACKNOWLEDGMENT

This book is essentially a study of urban and rural forces in Iowa politics during the years prior to 1964, when the state seemed to be experiencing an important political transition. Principal attention, therefore, is focused upon the role of history, voting behavior, interest groups, government officials, and political parties in the development of urban-rural conflict and cooperation. In the final chapter, however, an attempt is made to compare the major findings of this study with the results of similar investigations of other American states.

Support for this research was provided by fellowships from the Robert A. Taft Institute of Government and the Woodrow Wilson Fellowship Foundation and by a Horace A. Rackham grant from the University of Michigan.

Many people, of course, have contributed to this book in reporting valuable information about state politics and in offering helpful criticism and suggestions. I am particularly honored to join with many others in acknowledging my indebtedness to Professor V. O. Key, Jr., who played a major role in originating and guiding this research. In addition, appreciation is extended to H. Douglas Price, who provided needed encouragement and advice after Professor Key's death, and to Charles R. Adrian for his reading of the final manuscript. The author, however, assumes responsibility for all errors of fact and judgment.

Finally, I wish to express my gratitude for the inspiration provided by my parents, life-long Iowans, to whom this book is respectfully dedicated.

Harlan Hahn

CONTENTS

URBAN-RURAL CONFLICT

Chapter 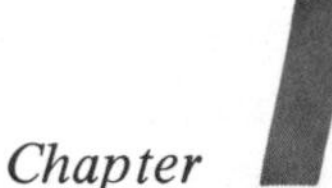

THE POLITICS OF GEOGRAPHY

Political conflict seldom develops in a vacuum. Although public debates and discussions usually focus on opposing policy objectives, contests between rival groups are promoted by a variety of conditions that divide the populations of nearly every state. In many areas, racial, religious, economic, and ethnic differences form important bases of struggles for political resources. Among the factors that encourage political strife, however, the division of urban and rural interests deserves serious attention.

Conflict between the residents of urban and rural areas long has been a prominent feature of American politics. Many of the early arguments for democracy were founded on a Jeffersonian belief in the sturdy independence of yeomen farmers and a corresponding depreciation of the tumultuous mobs that inhabited cities. As agrarian homesteads and small towns conquered previously unsettled regions, the nation generally retained its faith in the preeminence of rural America. Although growing numbers of urban dwellers had emerged by the dawn of the twentieth century to challenge rural influence, urban-rural conflict commonly has been portrayed as an uneven contest in which the cities have been forced to assume a subordinate role.

The claim of rural domination in American politics has been ascribed to several major sources. Many people, including some who live in cities, have subscribed to a belief in the superior virtues of rural life (see Gallup Poll Reports, March 23, 1966; May 5, 1968). Since this attitude provided a moral basis for the political ascendancy of rural areas, the perpetuation of rural control in many states was readily justified. Perhaps a more important contribution to rural power, however, has been the impact of tradition. Although America has been transformed from a predominantly rural to a predominantly urban nation, many politicians have found it less difficult to maintain existing arrangements that favored rural areas than to institute changes that reflected population shifts. Urban areas, for example, have been consistently underrepresented in legislative assemblies that have failed to reapportion their seats on the basis of population. The attainment of popular majorities by urban centers frequently has stimulated urban-rural friction over systems of legislative representation that discriminate against urban residents. Instead of ameliorating tensions, the tradition of rural dominance in legislatures as well as other areas of political activity has exacerbated urban-rural controversies.

Although urban-rural divisions contain numerous implications for the study of political conflict, prior research often has been limited to an attack on the evils of rural domination (Baker, 1955). The broader ramifications of the clash between opposing concentrations of population and political influence have been ignored in polemical denunciations of specific problems such as inequalities in congressional or legislative representation. Contrary to much discussion, urban-rural conflict has not been confined to the question of legislative apportionment. Conceivably, for example, the mutual suspicion of cities and rural areas has affected the course of two-party politics in America for a considerable number of years. Without departing from the realm of political analysis for the field of psychology, however, the study of urban-rural conflict should provide an increased appreciation of the role of geographical divisions in the distribution of political power. By shifting the focus away from the description of particular issues and from the merits of urban or rural positions, the research will seek to explore the importance of urban-rural cleavage as a factor in the formation of public policy and as an organizing concept for the investigation of political behavior.

In a widely quoted and respected analysis, Harold Lasswell (1958) once defined politics as the process of "who gets what, when, and how." As a natural corollary of the statement, political influence or power could be described as the ability to attain political objectives under desired circumstances. Although this definition of politics has served many accounts of political phenomena adequately, it also might be amended to include the term "where." Conflicts between opposing actors often have been affected not only by the techniques of acquiring political resources or by the immediacy with which they are obtained, but disputes also have been shaped by the location of the resources. A large portion of the time spent by most elected representatives, for example, is devoted to securing government programs or new public buildings for their constituencies. Since the needs of urban and rural districts often produce competitive demands, demographic factors may influence both the allocation of funds for policies such as agriculture and urban renewal and the communities in which they are to be located. Even more strikingly, political controversies are shaped by the environments in which they occur. Contests that are conducted in predominantly rural areas will differ from those in urban regions. In addition, the specification of political boundaries offers an example of the crucial role of location in the resolution of political conflicts. Urban communities might encounter substantial disadvantages in overwhelmingly rural governmental units, and vice-versa. Frequently, changes in political settings through the relocation of boundaries could alter radically both the nature and the results of governmental policies.

One of the most significant features of the environment of American politics is the federal system. It has become a virtual truism to note that in many respects American politics is not the politics of two national parties or of three branches of the federal government, but the politics of fifty states and hundreds of parties. An argument even might be propounded that each state constitutes a distinct "political culture" which most elected representatives must master before they can attain prominence in national politics. Consequently, this study will consider not only the political differences of urban and rural interests, but it also will examine urban-rural conflict within a particular American state.

In general, published studies of American state politics have tended to neglect the role of urban-rural cleavages. Research on state politics long has been concerned with the problem of party competitiveness. Yet, most prior analyses have devoted slight attention to the important effect of urban-rural differences on party competition. Although there may be some exceptions that prove the rule, students of political parties long have been familiar with the observation that most large cities in northern states tend to be predominantly Democratic while rural areas are inclined to be predominantly Republican. Perhaps more than coincidental is the fact that the major pivotal northern states with competitive party systems also contain large urban populations. As effective agents for political expression in state politics, parties are not only based on contrasting centers of electoral strength but they also might serve as the natural spokesmen for opposing urban and rural interests. By investigating the impact of urban-rural conflict on state party competition, significant progress might be made in the examination of the propositions that differences between parties mirror urban-rural disagreements and that urban-rural schisms may promote competitive party politics.

The relative neglect of the rural-urban dimension in political analyses has been fostered by a number of difficulties. Perhaps the major problems encountered by research on urban and rural cleavages concern the special characteristics associated with the two areas. Many urban communities probably have more attributes in common with rural areas than with other cities of the same size. Some small towns and farming areas also may acquire features usually found only in large cities. On the other hand, the distinctions that normally do exist between urban and rural areas are closely inter-related with other factors. Divisions between urban and rural regions generally can be ascribed to differences in occupational status, education, ethnic origins, income, and other social and economic variables. In other words, there commonly is no distinctive "urbanness" or "ruralness" that can be isolated as a source of conflict apart from other considerations. Furthermore, clashes between urban and rural interests often are clouded by urban residents who may sympathize with rural objectives and by rural inhabitants who may identify with the needs and aspirations of cities. Many of the policies and programs that stimulate urban-rural struggles may arouse ideological or partisan feelings that transcend urban-rural boundaries.

Such difficulties, however, need not impose an absolute prohibition on studies of urban-rural conflict. Despite the obstacles to a precise account of casual relationships, it is possible to gain an increased understanding and appreciation of one of the major configurations of political power in a state by examining urban and rural differences. Instead of seeking to disentangle urban or rural qualities from the host of other variables with which they are related, this analysis will treat the urban-rural dimension as an organizing concept and as a perspective for the analysis of state political systems. As a result, for example, urban-rural divisions in voting behavior will be examined by delineating the partisan choices of cities, small towns, and farming areas rather than by attempting to isolate the effect of urban or rural residence on electoral decisions exclusive of other factors such as income, education, ethnicity, etc. Similarly, the role of urban-rural cleavages in the activities of organized interests, governmental institutions, and the parties will be investigated by probing the contrasts between groups with urban and rural bases of support rather than by exploring the behavioral variations produced by geographical considerations apart from other organizational and socioeconomic components. Although the influences that normally are associated with urban and rural areas may tend to complement urban-rural distinctions, every effort will be made to minimize or eliminate the effects of extraneous forces in the research. The ability to remove other social and economic variables obviously would aid substantially in the clarification of urban-rural conflict, but the examination of state politics within the context of geographical differences also can contribute substantially to an understanding of political behavior by providing a framework or organizing principle for analysis and by specifying the probable impact of urban-rural tensions on other political groups and institutions. Since the mutual antagonisms and hostilities of urban and rural areas have been a major feature of American politics, there is a great need for descriptions and analyses of the nature and extent of urban-rural conflict. The purpose of this study, therefore, will be to explore the prevalence of urban-rural disagreements in the politics of an American state, the partisan features of urban and rural areas, the organization of urban and rural interests, the impact of urban-rural divisions on government structure, and the long-term potential and consequences of urban-rural disputes.

The analysis also will focus on politics in the state of Iowa. Although the political impact of urbanization will be explored through a comparative assessment of the politics of all fifty states in the final chapter, the research initially will attempt to isolate the major patterns of urban-rural conflict that have developed in Iowa. Emphasis also will be placed on the informal rather than the formal arrangement of authority in Iowa government (see Ross, 1957). The study is predicated on the assumption that "the very essence of state politics requires [coming to] grips with *whole state political systems* (Price, 1963:4)." By examining urban-rural cleavages within the confines of Iowa politics, it should be possible to gain an appreciation of the relation of urban-rural disagreements to state politics generally.

The late Senator Jonathan P. Dolliver (quoted in Ross, 1958:65) once predicted that "Iowa will go Democratic when Hell goes Methodist." For many years Republicans, who enjoyed a nearly total monopoly on public offices in the state, regarded this statement as fact rather than speculation. In the century following the end of the Civil War, there were only five Democratic governors; and only four Democrats represented Iowa in the United States Senate. Since a strong one-party tradition had been established, the prospect that Iowa voters might abandon their faith in the Republican party seemed remote.

However, there also were indications by 1964 that Iowa might have been undergoing a political transformation. Two of the five Democratic governors were elected in the decade from 1955 to 1965; and, in 1959, four of the eight members of the Iowa congressional delegation were Democrats. Similarly, in 1964, Democratic candidates for president, governor, and all state offices scored unprecedented victories in Iowa. After that election, Democrats also controlled six of the seven congressional seats and overwhelming majorities in both houses of the state legislature. While the Republican resurgence in subsequent elections demonstrated that most Iowa voters had not been converted to a new political faith overnight, there were indications that new forms of party competition based on an increasingly urban population may have been emerging in Iowa.

Although no universally accepted methods have been devised for classifying states which contain one-party or two-party systems, politics in Iowa for much of the period since the Civil War could be described as predominantly Republican. Only in an era that generally coincided with growing urbanization in Iowa has the Democratic party developed into an organization capable of providing an effective challenge for major Republican candidates. As a state that seems to have emerged from overwhelming Republican dominance, Iowa offers a rare opportunity to examine both the factors which support a tradition of one-partyism and the conditions which may promote a transition from one-party to competitive party politics.

Unlike many other states, Iowa has experienced few socioeconomic changes that might disrupt prevailing partisan loyalties. As one observer (Rogow, 1961:869) has concluded:

> Political, economic, and social homogeneity have been more characteristic of Iowa than of a large number of other states. . . . Even the usual rich-man-poor-man distinctions are more difficult to make; the extremes of wealth and poverty are not as visible in Iowa as elsewhere in the nation. There are no large numbers of recent immigrants to assimilate, and the state's non-white population has remained small. Political differences exist, but they exist less between the parties than within the parties, and within the parties they tend more to be personal or factional than ideological. At any given moment, therefore, the political mood of the state is likely to be conservative, cautious, and 'standpat.'

For many years, this description also could have been applied to urban-rural divisions. The slight differences that existed between urban and rural areas were muted by the overwhelmingly rural character of the population and by the attachment of the state to the Republican party. In the absence of other sources of conflict, however, there was little delay before urban-rural disagreements began to find political expression.

The relative lack of other struggles in Iowa politics permits a more intensive analysis of urban-rural friction than would be possible in most other states. Unlike many populous northern states, for example, Iowa does not contain numerous or diverse ethnic groups. The largest group of foreign born voters in Iowa are Germans. Since many German immigrants arrived in Iowa in the period between 1850 and 1860, they were strongly influenced by the same anti-slavery appeals that stimulated Republican loyalties among the majority of Iowa voters (Herriott, 1918). Similar motives apparently inspired Scandinavian and Dutch settlers to affiliate with the Republican party (Christensen, 1952). By 1860, therefore, the bulk of the foreign born population in Iowa was united in a common political tradition that was devoted to a single political party. Subsequently, the parties in Iowa have had little opportunity or necessity to issue appeals specifically designed to attract the support of ethnic groups. In 1940, for example, one historian (Parker, 1940:145) concluded that in Iowa "there has never been a candidate for either Governor or United States Senator with a German name."

Since most foreign born settlers originally located in rural areas, ethnic group differences have contributed relatively little to urban-rural cleavages in Iowa. Perhaps a rural environment reduced the self-conscious aspirations of ethnic minorities in politics. An immigrant farmer probably faced fewer social and economic obstacles than an urban factory worker. As a result, there was little incentive to turn to politics as a means of advancement. Furthermore, newly arrived Americans in a dispersed, homogeneous population such as that of Iowa probably clashed less frequently than in many other states. Since there have been other routes to recognition and prestige, ethnic competition and striving for political success has not been a marked feature of Iowa politics.

Ethnic and religious minorities seem to play their major role in Iowa politics by solidifying and reinforcing traditional partisan attachments. An analysis of the vote in all cities and towns in Iowa for which election returns were available from 1948 to 1964 revealed that the towns which had given the most consistent support to the Democratic and Republican parties, respectively, were each composed of one predominant ethnic and religious group.

The two communities that usually provide the most consistent support for the Democratic party are Carroll and Dubuque. Both of these towns are predominantly German and Irish Catholic settlements. Although the voting patterns of the two communities have mirrored the trends in the state generally, both towns gave Democratic candidates for president, governor, and congress a larger percentage of the vote than the remainder of the state in all elections from 1948 to 1964. Undoubtedly the overwhelmingly German and Irish Catholic composition of these communities has enabled them to withstand the currents that otherwise might have forced them completely into the Republican column. The only major exception to this rule developed during World War II when Dubuque voters were sufficiently aroused by the prosecution of the war against Germany to cast an unprecedented majority for the Republican candidate for president, Thomas E. Dewey. Aside from such rare occurrences, the homogeneous ethnic and religious nature of these communities reinforced rather than disrupted traditional partisan allegiances.

The same phenomenon is true for the Republican party. The two most consistently Republican communities in Iowa were the predominantly Dutch settlements of Orange City and Sioux Center in Sioux County. The voting patterns of these Republican towns experienced fewer fluctuations than the returns from the most consistently Democratic communities in Iowa. Throughout the decade from 1948 to 1958, the Republican vote in the two Sioux County communities was twenty or more percentage points higher than the state Republican vote. The fact that most of the residents of Orange City and Sioux Center are of Dutch descent and members of the Dutch Reformed Church undoubtedly accounts in part for their solidly Republican record. The impact of religious considerations on voting might be inferred from the election of 1960 when Orange

City residents cast 92.7 percent and 89.2 percent of the vote for the Republican candidates for president and governor, respectively, who were opposed by Catholic opponents. It is apparently easier to maintain traditional partisan attachments when a group is wedded to the dominant rather than the minority tradition in state politics.

Although the analysis of consistently Republican and Democratic communities might suggest that ethnic or religious differences also could intensify divisions between urban and rural voting patterns, a more comprehensive examination failed to confirm this proposition. While the four towns that provided the most consistent support for the two parties each were composed of a predominant ethnic group, they were among few settlements in which voting behavior may have been determined by ethnic or religious considerations. None of the remaining communities in the state approached the four communities in the unanimity of their partisan preferences or of their socio-religious composition. Furthermore, three of the four most consistently Republican and Democratic communities were small towns that might be expected to possess a high degree of social cohesion and conformity. Dubuque was the only urban area identified in which partisan allegiances seem to have been influenced by ethnic factors. Unlike many other states, Iowa politics has not been shaped by conflicts between large ethnic concentrations in urban areas and rural native born voters. Similarly, there has been almost no political competition or animosity between opposing ethnic groups. In a few communities, ethnic attachments apparently have strengthened party loyalties; but, in the overwhelming majority of both urban and rural places, ethnic affiliations seemed to have little impact on partisan electoral choices.

Iowa also lacks numerous other social and economic characteristics that have intensified urban-rural strife elsewhere in the nation. Although a popular image of the state as an exclusive preserve of corn and hogs is no longer accurate, Iowa traditionally has not contained a large working class population centered in metropolitan areas. For many years, the movement of organized labor has remained a small and relatively passive force in state politics. In addition, there have been few gigantic industrial corporations, after the peak of railroad power in the late nineteenth century, capable of dominating the economic and political life of the state. Since Iowa remained a comparatively rural state during much of the twentieth century, the political struggles that normally accompany gradual industrialization and urbanization reached Iowa considerably later than other regions of the country. The relatively sudden emergence of urban-rural disagreements in Iowa seems pressing and dramatic, but they probably have not been as intense as similar conflicts in large industrial states where the contrasting objectives of urban and rural areas are strengthened by vast economic and social differences. While the boundaries between urban and rural areas may reflect some socioeconomic singularities, the distinctions between cities and farming areas in Iowa generally have been less severe than in many states of the union.

In many respects, the industrialization and urbanization that has occurred in Iowa has represented a natural economic transition. Like many areas of the country, Iowa has experienced a consistent reduction in the population of farming areas and small towns and a corresponding growth in urban centers. The impact of this trend, however, has produced few of the economic and social dislocations that could develop among people who were unprepared and unequipped to cope with technological change. The population has been relatively well educated. Iowa has the highest literacy rate in the nation. The training and backgrounds of most residents of the state has enabled them to adjust with comparative ease to altered economic circumstances. Most of the young men and women who have migrated from farms, for example, have obtained white collar jobs or skilled work with small manufacturing firms locally or in the metropolitan areas of neighboring states. As a result of this movement into positions that offer at least moderate or middle class status, the development of urbanism and industrialization in Iowa generally has not produced the radical economic disparities that frequently have afflicted other regions with relatively low rates of literacy and educational attainment.

Although the incomes of most Iowa residents fail to reveal outstanding inequalities in economic status, the occupational characteristics of the state generally can be separated along urban-rural lines. Most of the work performed in rural areas is directly or indirectly connected with agriculture. In addition to farming, much of the work performed even in small towns is devoted to serving agricultural interests. On the other hand, urban centers commonly are identified by the criteria of population and substantial employment in commerce or manufacturing. Although many city residents are engaged in tasks that eventually benefit rural inhabitants, occupational differences provide at least one means of distinguishing urban and rural places.

A major difficulty in the study of urban-rural politics concerns the problem of definitions. Clearly, the designation of urban and rural areas varies greatly by time and by region. A community of 45,000 people might be considered a relatively large city in a small, sparsely inhabited state or in an early period of history; but it probably would be termed a small or medium-sized town in a modern or densely populated state. Concepts of urban and rural areas have changed as settlements have expanded and populations have increased. Furthermore, few recognized criteria have been of value in establishing the demarcation between urban and rural regions. The United States Bureau of the Census definition of urban places as settlements of 2,500 or more inhabitants cannot be considered very helpful, for example, because it places towns of 3,000 in the same category as cities of 3,000,000. Such a classification system principally separates the residents of any settled community and people who live in farms or hamlets rather than major population centers and sparsely populated areas. Adopting a distinction between farms and communities rather than between populous concentrations and small habitations clearly would inflate the urban category. Even in Iowa, farmers constitute a much smaller proportion of the voting population than is commonly supposed. As early as 1885, for example, 36.9 percent of the eligible voters in the state lived in incorporated towns. Within another decade, the total number of people residing in towns had surpassed the number living on farms. Since farms probably have not formed the principal sources of rural strength in Iowa politics, the focus of this investigation of urban-rural conflict will not be limited to the political differences between farmers and the residents of small towns.

For the purposes of this study the occupational and social characteristics of Iowa communities were examined to establish a population of 10,000 as a general dividing line between urban and rural areas. Cities in Iowa of 10,000 or more usually contained at least one major manufacturing plant, while towns below that mark normally lacked an industrial plant that provided employment for a sizeable proportion of the residents (Hahn, 1963). By this definition, 61.9 percent of the population of Iowa could be considered rural according to the 1960 census. On the other hand, 53.1 percent of the population of the state was concentrated in the 20 counties containing cities larger than 10,000 in which a substantial share of the population was urban. Within the 20 largest counties, the residents of cities of 10,000 or more population constituted 71.8 percent of the county inhabitants.

Although the level of 10,000 population seems to provide an appropriate demarcation between urban and rural areas, the analysis of state politics should not be confined by iron-clad distinctions. At various times and in different localities significant urban-rural cleavages have been evident at lines either above or below that figure. Groups that have engaged in urban-rural contests on both sides have included the residents of different sized communities. In addition, the alignments between urban and rural forces have shifted according to the issue and the circumstances. On some occasions, for example, voters in large cities and small towns have joined in opposition to farming interests; but often farm and small town residents have merged to oppose city dwellers. While a standard division may aid in clarifying the meaning of urban and rural areas, therefore, the examination of major segments of the population must be sufficiently sensitive to detect important urban-rural differences at various population levels and locations.

The definition of urban and rural places in Iowa also is complicated by the unusual geographical features of the state. Unlike many states, Iowa does not contain a major metropolitan center. Instead, the bulk of the urban population is contained in a number of medium-sized cities that are scattered throughout the territory. Des Moines, the largest city in Iowa and the state capital, had a population of only 208,982 in 1960. The non-metropolitan character of Iowa probably has reduced the potential for urban-rural conflict. Perhaps in states that contain a major metropolitan area such as New York, Chicago, and Boston, or in other urbanized states such as Michigan, Pennsylvania, Ohio, or Wisconsin, there are increased opportunities for the promotion of urban-rural rivalries and for the domination of politics by either the major population concentrations or the remainder of the state. On the other hand, the absence of a large metropolitan complex would seem likely to prevent the development of urban solidarity and retard the formation of an urban coalition in state politics. Furthermore, Iowa contains few suburban developments, which frequently add another refinement to urban-rural classifications elsewhere. In 1960, only 6.7 percent of the residents of Standard Metropolitan Statistical Areas were located in urbanized places of less than 10,000 population (U.S. Bureau of the Census, 1963). The relative lack of suburbs could exacerbate conflict by removing a group that frequently forms a "buffer zone" between urban and rural areas and that occasionally may ally with conservative rural interests, despite its urban orientation. In Iowa, however, the absence of both metropolitan and suburban developments probably has contributed to political constancy by eliminating the potentially disruptive effects of new political interests and demands.

In addition, the relative stability of Iowa politics clearly has been promoted by a number of other attributes. Iowa, for example, contains the largest proportion of people over 65 years of age in the country. Since older persons normally are particularly reluctant to depart from established traditions and conventions, the relatively advanced age of the population may have tended to reinforce political continuity in the state. Perhaps even more politically relevant, however, is the fact that Iowa contributed more men per capita to the Union army in the Civil War than any other state. As a result, partisan loyalties formed during that period probably were perpetuated more effectively in Iowa than in many other parts of the North. The political inheritance of many Iowans was molded in the mid-nineteenth century; and few subsequent social or economic upheavals, aside from urban migration, have disturbed this legacy.

Perhaps the most serious threat to prevailing political traditions in Iowa has been the relatively rapid development of urban areas during the era after World War II. In this period, there has been a relatively steady decline in the proportion of people living on farms or in small villages and a concomitant increase in the urban share of the state population. Even more indicative of growing urbanization have been the shifts in the economy of the state. As late as 1949, the value of agricultural products exceeded the value of industrial production in Iowa. In little more than a decade, however, the wealth contributed to the state by manufacturing was greater than that derived from agriculture by more than a two-to-one margin. While the value of agricultural production remained relatively constant, the value of industrial output more than doubled during the period (Schmidhauser, 1963:1-2). The relatively sharp alterations in the urban and rural characteristics of the state had a major impact upon political configurations in Iowa.

Although changes in the urban-rural composition of a state may have continuing if not enduring consequences, no investigation of state politics can include appraisals of effects that may extend far into the future. Moreover, certain eras might be more critical in the development of state political processes than others. Perhaps the most fascinating as well as the most significant period in urban-rural controversies is the time encompassed by the transition from a predominantly rural to a predominantly urban political environment. In Iowa, most of the forces that played a crucial role in urban-rural disputes seemed to have emerged by 1964. As a result, this study of Iowa politics is limited to the period between the state's admission to the union and 1964. By that time, the historical trends that carried the seeds of both urban-rural conflict and consensus seem to have reached fruition.

In many respects, Iowa represents a state that has entered a period of rural-urban transition. Conflicts between urban and rural areas probably are more divisive and more visible in many other sections of the country. To some residents of more populous states, therefore, both the definition of urban areas and the emphasis on urban-rural tensions in Iowa might appear to be misplaced. Perhaps the bitter struggles over political resources between New York and Albany or between Chicago and Springfield could provide a clearer example of the potential intensity of urban-rural divisions than the conflicts that have emerged in Iowa.

Yet, the focus on a state in a stage of transitional development can provide valuable information about the process by which urbanization has influenced politics elsewhere. During the years since the Civil War, America has changed from an overwhelmingly agrarian nation to a predominantly urban society. This experience undoubtedly has had a major impact upon the course of political history. In addition, the emergence of newly independent countries in Africa and Asia has demonstrated that a close relationship exists between rates of urbanization and political development. Since it has been nearly impossible to retrace the stages through which many American states have passed in reaching their present levels of political maturity, however, the investigation of a state at a crucial phase of urbanization may be a necessary precondition for a broad evaluation of political trends. Detailed information about a state that is undergoing the transition from rural to urban characteristics, therefore, could aid in understanding both the origins and the future development of political conflict in other jurisdictions.

The objectives of this study are not only to identify the subtle manifestations of urban-rural differences in Iowa but also to explore the impact of urbanization upon the aggregate features of state politics generally. In many respects, the former purpose may be a critical prerequisite to the latter goal. Yet, the findings of this research eventually must be compared with the major contours of politics in other states. The usefulness of urban-rural distinctions as an organizing concept in the analysis of state politics can be measured not only by its ability to contribute to the explanation of politics within a single state but also by its value in developing a comprehensive interpretation of American state politics. In the final chapter, therefore, a comparative investigation of politics in all fifty states is attempted both to provide a means of cross-checking the major outlines of Iowa politics and to probe the theoretical utility of urbanization in the study of American state politics.

This comparison, however, necessitated some modifications in both the scope and the approach of this study. Although a population of 10,000 seemed to form an appropriate level for identifying urban areas in Iowa, for example, it did not appear to be a feasible definition for states in more advanced phases of urbanization. In addition, some provision had to be made for the fact that Iowa does not include a major metropolitan area. As a result, both the total proportion of persons living in communities of 10,000 or more and the presence of a dominant city of more than 250,000 or 500,000 population were employed in the general examination of state politics. The absence of a large metropolis, however, seemed to offer a partial advantage rather than a limitation in the use of Iowa as a benchmark for the comparative study of American states. Whereas metropolitan centers are found in all of the most populous states in the country, nearly half of the states have not contained a city of either 250,000 or 500,000 population. Perhaps many of those states have exhibited a closer resemblance to the political characteristics of a state such as Iowa than to other sections of the country that have received more attention. Politics in Iowa probably has displayed similarities with a larger number of states than politics in New York or California. As a state that has entered a transitional phase of urbanization, therefore, Iowa seemed to provide an adequate and interesting basis for the study of American state politics.

Although urban-rural cleavages often have stimulated important conflicts, many states have developed political conditions that may reduce rather than promote controversy. Differences between urban and rural areas frequently have been the foundations of major partnerships as well as the sources of intense battles. In fact, the predominant political heritage of Iowa has tended to discourage urban-rural contests rather than to provoke them. The history of Iowa politics has encompassed many joint endeavors between urban and rural interests. As a result, attention must be devoted to both urban-rural conflict and cooperation.

Since urban centers were not sufficiently large or numerous to provide a target for antagonism in the early history of Iowa, the sources of modern urban-rural tensions developed during an era in which disagreements between farmers and the residents of small towns or between agricultural and business interests constituted the principal forms of urban-rural conflict. While the schisms between country and village areas frequently were muted, they seem to have contributed in somewhat unusual and significant ways to subsequent struggles between cities and rural areas. To some extent, differences involving small towns and farming areas as well as major urban concentrations have been reflected in partisan trends within the state. Much discussion of the so-called "farm vote" based on election returns from rural counties, for example, has concealed sharp distinctions between the voting behavior of farmers and the residents of small towns.

The political positions of interest group spokesmen and elected officials also have sometimes failed to disclose strong urban-rural distinctions. On occasion, urban interests and office holders have appeared to act in concert with rural forces, and vice-versa. A careful examination of those activities, however, seemed to suggest relatively consistent explanations for the behavior of lobbyists and government representatives. Furthermore, as the distinctions between urban and rural areas have sharpened, disagreements between the two groups also have been intensified.

Similarly, urban-rural characteristics have played an increasingly prominent role within political parties. Party efforts have been shaped not only by the environments in which they are located but also by specific traits that seem to be associated with partisan groups in urban and rural areas. Although no conclusive evidence has been offered for the proposition that large urban centers are necessary for effective partisan competition, contests between political parties as well as other forms of conflict probably have been molded in part by population distributions. The geographic concentration or dispersion of people, therefore, has been widely regarded as a significant feature of state politics.

REFERENCES

BAKER, GORDON E. (1955) Rural vs. Urban Political Power. Garden City: Doubleday & Co.

CHRISTENSEN, THOMAS PETER. (1952) A History of the Danes in Iowa. Solvang, California: Dansk Folkesamfund.

HAHN, HARLAN. (1963) "Reapportionment, the people, and the courts." Iowa Business Digest 34 (August): 19-22.

HERRIOTT, F. I. (1918) "A neglected factor in the anti-slavery triumph in Iowa in 1854." Deutsch-Amerkanische Geschichtsblatter 18 (Jahrgang): 1-170.

LASSWELL, HAROLD D. (1955) Politics: Who Gets What, When, How. New York: Meridian Books.

PARKER, GEORGE F. (1940) Iowa Pioneer Foundations. Iowa City: State Historical Society of Iowa.

PRICE, HUGH DOUGLAS. (1963) "Comparative analysis in state and local politics: potential and problems." Unpublished paper presented at the annual meeting of the American Political Science Association, New York, September 6.

ROGOW, ARNOLD A. (1961) "The loyalty oath issue in Iowa, 1951." American Political Science Review 55 (December): 861-869.

ROSS, RUSSELL. (1957) The Government and Administration of Iowa. New York: Thomas Y. Crowell Co.

ROSS, THOMAS R. (1958) Jonathan Prentiss Dolliver. Iowa City: State Historical Society of Iowa.

SCHMIDHAUSER, JOHN R. (1963) Iowa's Campaign for a Constitutional Convention in 1960. New York: McGraw-Hill Book Co.

U.S. BUREAU OF THE CENSUS. (1963) U.S. Census of Population 1960: Characteristics of the Population, vol 1, part 17. Washington: Government Printing Office.

THE REPUBLICAN TRADITION

Although political conflicts seldom are produced instantaneously, the history of struggles for governmental resources can assume a number of forms. Often, for example, battles between social or economic groups are inspired and perpetuated by a succession of specific incidents involving the principal contestants and by recurring rivalries and animosities. Other forms of disagreement, however, may evolve gradually and almost imperceptibly from the political climate or environment of an area rather than from a series of direct confrontations. Since sources of contention are difficult to isolate in such circumstances, attention must be shifted from outstanding political clashes to heritages that may encompass as well as divide a population.

The political customs and habits of a people normally have an influence that outlives the events that gave them birth. While traditions have their origin in the common experience of a homogeneous population, they do not always begin in the veiled obscurity of early history. Time is a relative concept; and a state that is slightly more than a century old is as likely to develop a legacy as a state that has survived for a millennium. Since political values usually continue through several generations, traditions in politics must be viewed as dynamic rather than static developments. Thus, tradition is considered here as an on-going process and as a carrier of attitudes as well as of values and habits.

In Iowa, the history of urban-rural tensions generally has conformed to this concept of tradition. Although the state was populated at a time when sizeable urban centers were rare, the history of Iowa politics has experienced a comparatively short development. Yet, within this time, urban-rural differences have emerged as a major form of political conflict less as a result of particular incidents than through the common experiences of the population. Political traditions have carried the seeds of both urban-rural conflict and cooperation. Only through the careful investigation of general trends in the state, therefore, will it be possible to identify the strands of history that have given birth to modern urban-rural cleavages.

As agents for the transmission of political symbols and allegiances, organized parties clearly must occupy a significant place in the study of political history or traditions. Party fidelity commonly has been stimulated by poignant memories and by the natural human impulse to shun departures from conventions as well as by the activities of parties themselves. Since partisan traditions in America have supported efforts to appeal to both urban and rural residents, party affiliations have tended to transcend urban-rural boundaries. Despite the efforts of party leaders, however, struggles occasionally have emerged to disrupt prevailing loyalties and to arouse an underlying heritage of urban-rural tension. As in many areas of political analysis, therefore, political parties should be accorded a prominent role in the examination of the history of state politics and the background of urban-rural differences.

In large measure, the overwhelming success of the Republican party in Iowa might be ascribed to the force of tradition. In part, this development may reflect what Meredith Willson (1962:148) has termed "Iowa Stubbornness." As the residents of a state that has been widely noted for its firm adherence to conventional values and established routines, Iowans perhaps have been more susceptible to the influence of traditions than the citizens of many other states. But the determined defense of personal values and beliefs has not been solely responsible for political stability. The political consciousness of Iowa was awakened by a struggle that had enduring consequences throughout the nation. The issues and events that precipitated the Civil War, along with the war itself, produced a political heritage that seldom has been overcome in the subsequent history of the state. The intense loyalties stirred by the Civil War probably were major ingredients in Iowa's Republican legacy.

Although tradition has provided a major advantage for the Republican party, Iowa was originally a Democratic state. From the admission of Iowa to the union in 1847 until 1854, Democrats controlled most major state and local offices. During the agitation preceding the Civil War, however, Republican roots were planted in the Iowa soil.

The early Democratic party in Iowa was "Southern in its antecedents and Jacksonian in its declarations" (Cole, 1921:105). During the debate over the entrance of Iowa to the union, George Wallace Jones, who was later to become one of the first two Democratic Senators from the state, argued that the majority of Iowa settlers were southern in origin and sympathy. In reply, Senator John C. Calhoun predicted (quoted in Cole, 1921:143), "Wait until western Ohio, New York, and New England shall pour their populations into that section, and you will see Iowa grow some day to be the strongest abolition state in the Union." Within a decade, it was becoming evident that Calhoun's prophecy had been substantially accurate.

In 1854, James W. Grimes, a young lawyer from southeast Iowa, accepted the Whig nomination for governor and launched what was probably the first aggressive political campaign in the state. Although Grimes championed prohibition and other measures, his opposition to the extension of slavery was the crucial issue in the campaign. Complacent Democratic leaders failed to examine census statistics which showed that settlers from the Northwest and New England had replaced southerners as the majority group in the state. At this time "Iowa began and remained a state whose roots were in the New England and the Middle Atlantic States" (Bergmann, 1956:135). The victory of Grimes represented a turning point in both the politics and characteristics of Iowa. As the agitation over slavery increased, the Democrats lost favor while opposition groups gained support among Iowa voters.

The Republican party in Iowa was born at a meeting in Iowa City on February 22, 1856. Although Grimes provided common leadership for the new group, the formation of the party involved the delicate task of uniting diverse and conflicting factions of Free Soilers, abolitionists, Know-Nothings, and former Whigs (Younger, 1955:78). Significantly, the platform adopted at the Iowa City meeting was "devoted exclusively to the slavery question" (Millsap, 1950:99). Although the delegates found it difficult, if not impossible, to agree on other policies, the slavery issue was sufficiently absorbing to provide the basis for a party organization that sought to ensure its perpetuation in power.

Meanwhile, even before the formation of the new party, the groups opposed to the Democrats enacted a series of organizational and institutional changes to solidify their position. Boundaries of congressional constituencies, judicial districts, and countries were altered or redrawn; and preparations were made for a new constitution to replace the 1846 document, drafted largely by Democratic delegates, that prohibited the establishment of banking institutions in the state. Unlike the earlier constitutional convention, the majority of the delegates to the meeting that framed the 1857 constitution were natives of northern states. The new constitution, which is still the basic framework for Iowa government, was ratified "by the close vote of 40,311 to 38,681–a majority of only 1,630. This was practically a strict party vote, and stands in about the same ratio as the later vote in the regular state election of the year 1857" (Clark, 1911:10). Within a few years, changes in the population and sympathies of Iowa had transformed the character of government and party competition in the state.

The Republicans were perhaps uniquely fitted to capitalize on the events of this era. Slavery was the one issue upon which the party could unite. The intense emotions aroused by this issue forced all other considerations into the background and prevented the party from fragmenting or quarreling over other questions. Similarly, as a young party, the Republicans were not plagued by the problem of consistency. They were not associated with positions on other issues nor had they taken stands in the past that conflicted with their statements on slavery. When the war finally erupted, the Republicans were in the enviable stance of spearheading a cause with which the entire state was associated.

Shortly after the opening shots on Fort Sumter, Governor Samuel J. Kirkwood (quoted in Cole, 1921:334) wrote President Lincoln, "Ten days ago there were two parties in Iowa. Now, there is only one, and that one for the constitution and the Union unconditionally." Perhaps more succinctly than any other comment, Kirkwood's observation illuminated the origins of the prevailing Republican tradition in Iowa. Since the Republican party had been closely identified with the issues and events that precipitated the war, it naturally became associated with the prosecution of the war. The Republicans were indelibly imprinted on the minds of Iowa voters as the only party that opposed a commonly hated foe. The outbreak of the war solidified and fixed this impression in the memories of men for generations.

As the war progressed, opposition to a common enemy produced a unity that was intolerant of deviant behavior. Gradually, in Iowa "the terms Democrat and traitor became synonymous" (Clark, 1911:139). While several of the Democratic leaders supported the war, many Democrats also were prominent "Copperheads," or supporters of the southern cause. A Democratic party meeting on August 1, 1863, sparked the so-called "Skunk River War" when a Copperhead orator was killed in a burst of gunfire. An attempt to form a separate army to avenge his death was met by the Republican Governor who issued harsh threats and arms to quell the disturbance. As a result the so-called "Skunk River Army" disbanded, and the murderers were never brought to trial (Cole, 1921:361-362). Such events heightened the popular identification of the Democrats and the Confederacy. Consequently, the Republican party became the only acceptable agent for political expression in Iowa, and the Democrats seemed "doomed to be a helpless minority for years to come" (Clark, 1911:8).

"The Civil War made the Democratic party the party of the South and the Republican party, the party of the North. . . . These loyalties were long sustained by the bloody shirt and the rebel yell" (Key, 1958:254). But Iowa may have been affected more by Civil War memories than many other states. In fact, the practice of waving the "bloody shirt" as a reminder of Republican sympathies during the war was originated in Iowa by General James B.

Weaver, who eventually gained national prominence as the Populist party candidate for president in 1892 (Haynes, 1919:24-25). As a state that underwent agitation over the Civil War during its formative years, Iowa was especially susceptible to the influences of traditional Republicanism.

The partisan preferences inspired by the war experience remained visible in Iowa politics for many years. In 1908, for example, when the internecine conflict over "progressivism" was at its peak in the Republican party, the "standpat" candidate, Major John F. Lacey, devoted his campaign leaflets almost exclusively to a description of his Civil War record. Beneath a picture of the candidate in his army uniform was a list of each of his promotions during the war along with the statement that he had "voted the Republican ticket exclusively from 1862 to the present time." Despite the pressing nature of other controversies in this second primary election in Iowa, the Civil War apparently remained a salient campaign issue for at least one of the candidates nearly fifty years after it had started.

In the years immediately following the Civil War, Republicans had little need for a formal organization to bolster their electoral majorities. As the attrition of the years began to thin the ranks of former Union soldiers and to cloud memories of the war, however, Civil War veterans intensified their organizational and political efforts. The permanent Iowa Department of the Grand Army of the Republic finally was established in 1879. "Within five years 16,500 'old soldiers' had enrolled. Practically all were Republicans, and many were intensely interested in keeping alive the memories of the Civil War" (Ross, 1958:34). Although Union veterans held a virtual monopoly on public offices in the state for many years, the G.A.R. probably contributed more to the Republicans than the soldiers required of the party. As one former prominent Iowa politician observed, the G.A.R. "maintained their power for only about fifty years. But they planted a political habit and a mood of thought." By constantly reviving and reinforcing potent remembrances of the war, the G.A.R. became a major bulwark for perpetuation of the Republican tradition when time threatened to eclipse the original impact of the war.

The organizational work of the G.A.R. as well as social pressures engendered by the war tended not only to support the Republican party but also to subdue latent antagonisms that might disrupt prevailing political values. Since voters for many years were preoccupied by the problems of the war, the prospects that intense political conflict would emerge as a major feature of Iowa politics seemed rather remote. Once major factions had united to oppose the extension of slavery and the Democrats had been discredited, the Republican hegemony appeared indestructible. With the exception of a small minority of Copperheads and pre-war Democrats, Iowa voters were united by a common devotion to the union cause that yielded few differences in the political preferences of major segments of the population.

In 1868, Iowa held two elections that permitted a careful investigation of groups supporting both the Republican party and one of the principal issues that precipitated the war. At the same time, voters were given the opportunity to cast their ballots for General Ulysses S. Grant, the Republican nominee for president, and on five referendum proposals to repeal discriminatory clauses of the Iowa constitution concerning suffrage and legal equality for Negroes. Unlike at least ten other northern states that defeated proposals for Negro suffrage, Iowa gave more than two-thirds of its total vote to Negro rights as well as to the candidacy of General Grant. An analysis (Dykstra and Hahn, 1968:202-215) of 437 township election returns revealed that farm areas and towns of 2,500 or more supported Grant and Negro suffrage by identical margins of 62 percent and 56 percent, respectively. Villages of less than 2,500 population, on the other hand, favored Grant by 65 percent and Negro suffrage by 61 percent. The principal opposition to Negro suffrage was concentrated in predominantly Democratic townships and counties. After Appomattox, voting patterns on the issues that fomented the war and partisan elections were highly interrelated. While villages were slightly more favorable to Negro suffrage and the Republican presidential candidate than small towns or farming areas, conflict between urban and rural areas had not emerged by the end of the war as a prominent feature of Iowa politics.

During the era that was dominated by the Civil War, Iowa experienced a major influx of population. Between 1860 and 1870, the population of the United States increased 22.6 percent; in Iowa, it increased 76.9 percent. Most of the migrants, who were from the Northeast, were drawn into the political life of the state at a time when there seemed to be only one proper partisan alternative. Under these formative circumstances, it was not surprising that the settlers and many of their descendants continued to support the Republican party.

Even more striking is the fact that most of the settlements that flourished between 1860 and 1870 were devoted almost exclusively to agriculture. "The number of new farms established during this period was 55,129, or an increase of 90% in Iowa" (Haynes, 1916:199). The basis for social, economic, or political cleavages along urban-rural lines did not exist because there were no major urban concentrations and few residents were engaged in nonagricultural pursuits.

The Republican party also made a significant effort to attract farmers in Iowa and other states through the adoption of the Homestead Act. A comparison of counties that were settled before and after the passage of the Homestead Act (Blackhurst, 1959) indicates a high association between Republican voting and settlement after the passage of the act that persisted until the election of 1892. Although the relationship between support for the Republican candidates and settlement after the Homestead Act was influenced by the proportion of settlers born in the South, the act undoubtedly provided the Republican party with at least a strong symbolic appeal for farmers in Iowa.

The approval that the Homestead Act and the Republican party received from rural voters, however, was clouded by the fact that few Iowa farmers benefited directly from the act. By the time the Homestead Act became effective, the bulk of the available farming land in Iowa had been purchased by railroad companies or land speculators either from the government or from private owners (Gates, 1964). The provisions of the act were a disappointing delusion for many Iowa farmers who had expected to obtain land at a nominal cost from the government. Naturally the necessity of acquiring property from railroad and land companies spawned antagonisms between farmers and the business concerns that reaped a substantial profit on the venture. Thus, the seeds were sown for a conflict between farmers and business interests that was destined to have profound political repercussions during much of Iowa history.

The Republican party had scarcely been established in Iowa when business interests began to employ political means to achieve their goals. Immediately after the election of Lincoln, a group of Republican businessmen and office-seekers, which has come to be known as "Dodge and Co.," rented a house in Washington, D.C., where they attempted to wrest railroad legislation, contracts for trade with the Indians, and other prizes from the new administration (Sage, 1956:43). The undisputed leader of the group was General Grenville M. Dodge, a railroad promoter who later received a commission during the Civil War. Included also were Dodge's brother, Nathan; John A. Kasson, who had guided the Republican party through the campaign of 1860; the newly elected Republican state chairman, Herbert M. ("Hub") Hoxie of Des Moines; and William B. Allison, a young Republican lawyer from Dubuque who aspired to be the United States District Attorney for Iowa. The political organization in Iowa that Dodge and his friends built with business favors and federal patronage gradually became the dominant faction in the Republican party.

The major opposition to "Dodge and Co." arose from the supporters of Senator James Harlan, who had joined with James Grimes in breaking the Democratic monopoly prior to the war. After resigning his seat in the United States Senate, Harlan reconsidered and sought the seat that he had vacated. Since Governor Kirkwood also had announced his Senatorial candidacy, Harlan was opposed by Dodge and Kirkwood. In the Republican legislative caucus of 1866, however, Harlan secured the votes for the full Senate term, leaving Kirkwood with the shorter unexpired term.

Meanwhile, the tides of Radical Republicanism swept Iowa. Kirkwood spearheaded the Radical drive, and Dodge seized the opportunity to purge John A. Kasson, who had failed to espouse the Radical cause in Congress. In his only bid for major elective office, Dodge defeated Kasson for the Republican congressional nomination (Younger, 1955:184).

Similarly, in 1870, "Dodge and Co." were active in another contest when Grimes was ousted from the Senate because of his failure to support the impeachment of President Andrew Johnson. The nominee that Dodge and his associates finally chose was Congressman William B. Allison. Dodge assumed absolute command of Allison's campaign and sought to advance his candidacy through the distribution of "business favors, loans on easy terms, [and] promises of patronage" (Sage, 1956:93). Although Allison was defeated by George G. Wright, who was supported by the Harlan faction, there were significant indications that "Dodge was emerging as political boss of the state and as a powerful influence in national affairs" (Younger, 1955:209).

In December, 1871, "Dodge and Co." attracted a significant new ally when James S. ("Ret") Clarkson purchased a controlling interest in the forceful Des Moines publication, the *Iowa State Register*, from his father, Coker F. Clarkson, who had supported Harlan. "Ret" Clarkson shifted the position of the newspaper abruptly and joined the Dodge-Allison wing of the Republican party. As state Republican chairman, national committeeman, and assistant Postmaster General, Clarkson later employed a sizeable amount of patronage to maintain the dominance of "Dodge and Co." in Iowa politics.

The ultimate test of strength for the two major factions came in 1872 when "Dodge and Co." again supported the candidacy of Allison for the Senate seat that Harlan occupied. In a bitterly fought intraparty contest, Allison defeated Harlan and established a position in the United States Senate that he was to maintain for Dodge and his railroad interests during the remainder of the century. "Behind the cooperative Allison stood General Dodge, the expert wirepuller and . . . the head of the Dodge-Allison-Clarkson wing which was to dominate Iowa politics until 1908. Republican politics in Iowa had become in large measure the politics of railroads and other business" (Younger, 1955:242).

The extensive railroad commitments that Dodge controlled finally forced him to yield his leadership of the dominant Republican faction in Iowa to other men (Woodward, 1956:78-104). But the basic purpose of the Dodge-

Allison-Clarkson group remained dedicated to the promotion of railroad and other business interests. Although railroad companies received substantial concessions of land and subsidies from the state during this period, several efforts to adopt laws regulating the railroads were defeated in the state legislature, which was controlled by the Dodge-Allison-Clarkson faction. Nor could the leaders of "Dodge and Co." completely restrain their desires to participate in the spoils of politics. Both Dodge and Allison were prominent figures in the infamous *Credit Mobilier* scandal. In 1870, General Dodge also obtained the leadership of the commission to supervise construction of the Iowa state capitol. Within three years, it was discovered that the foundation stones were crumbling and had to be replaced at a cost of approximately $52,000 (Younger, 1955:244).

The extensive advantages enjoyed by business and railroads during the reign of the Dodge-Allison-Clarkson faction provoked resentment among other groups in the state. Particularly farmers, who were forced to rely upon railroads as the principal means of transporting their products to market, began to challenge the political rewards that "Dodge and Co." bestowed upon the railroads. The first major signs of agrarian protest in Iowa found their political expression in the Anti-Monopoly party developing from the Grange movement that flourished in Iowa at the time the Dodge-Allison-Clarkson wing was consolidating its strength in the Republican party.

Significantly, Clarkson and several other leaders of "Dodge and Co." apparently attempted to foster the new movement of farmers in the belief that they could control it (Younger, 1955:249,422). By enlisting the support of the recent members of the Grange, the Dodge-Allison-Clarkson group sought to prevent the defection of farmers to the Harlan faction or to the Democratic party. In part, their efforts may have been successful. Although the election of

Governor Cyrus Carpenter in 1871 generally has been ascribed to the support that he received from fellow farmers and members of the Grange, his candidacy apparently was launched by the Dodge-Allison-Clarkson faction. On January 7, 1871, George Tichenor, Dodge's principal political lieutenant and Republican state chairman, wrote to Dodge urging him to obtain a candidate for Governor. Dodge merely forwarded the letter to Allison with the note (quoted in Sage, 1956:107) "Suppose we put forth C. C. Carpenter of 6th Dist. We have no time to lose." As a result, Carpenter was nominated and elected.

Within a few years, however, the Anti-Monopoly movement became so large that it could not be controlled by either the Dodge faction or the Republican party. At the Anti-Monopoly state convention of 1873, a fusion of the Anti-Monopoly and Democratic parties was accomplished when John P. Irish, Democratic state chairman, proclaimed (quoted in Haynes, 1916:70) that "the Democratic party is dead." Although former Democrats assumed the leadership of the merged party, the Anti-Monopoly organization was essentially a rural expression of protest against the Republican party and its dominant faction. "The real issue behind the Anti-Monopoly movement in Iowa from 1873 to 1875 was the recognition of the right of the State to regulate rates in the interest of the people" (Haynes, 1916:82). As a prominent Granger, Governor Carpenter was not seriously threatened in his bid for reelection; but, at a local level, the movement initiated by dissatisfied farmers provided at least a temporary defeat for the Dodge-Allison-Clarkson faction and the Republican party.

In the election of 1873, the combined strength of Anti-Monopolist farmers and Democrats was sufficient to carry one congressional district and to elect a majority in the Iowa legislature. As a result the legislature finally passed the so-called "Granger Law" of 1875 to establish maximum rates for the transportation of freight. When the constitutionality of the law was upheld, many farmers and Grange members assumed that they had defeated the railroad interests in Iowa politics at last. At the same time, Democrats were attracted back to their own party by the national campaign of 1876 and the Anti-Monopoly party collapsed through neglect.

The supposition of farmers that the "Granger Law" would correct the abuses of the railroads was not confirmed, however. Within a few years, railroad interests succeeded in persuading the legislature to repeal the law and to substitute "an act of 1878 that created a commission pretty much controlled by the railroads" (Sage, 1956:160). Furthermore, after the demise of the Anti-Monopoly movement, the railroad interests strengthened their grasp on the Republican party in Iowa through the creation of an elaborate network of political influence.

While the bulk of the opposition to "Dodge and Co." was concentrated in rural areas of Iowa, principal support for the railroad interests tended to develop in small towns and urban areas. To attract the favor of voters and legislators, the railroads relied primarily upon prominent leaders in each major community. A doctor and a lawyer in every county seat town in Iowa were retained by the railroads to perform whatever services might be needed. The two professional men also were given free passes on the railroads which they could distribute to prominent citizens and businessmen in the community. As some of the most prestigious residents of the towns, the recipients of the free railroad passes frequently dominated local Republican caucuses and conventions (Bray, 1957:15-17). Significantly, two major groups were excluded from the favors bestowed by the railroads: Democrats and farmers who lived outside the towns. The favoritism practiced by the railroads naturally intensified the tension between rural settlers and the urban Republicans who supported railroad interests. Since a free pass was highly valued when railroads were the principal means of transportation, however, the groups opposed to the railroad interests found themselves at a considerable disadvantage in their efforts to obtain political leadership and support. Lacking both resources and prestige, Democrats and agrarian political movements were rendered impotent by the same methods that had been used to capture the Republican party.

Two of the most adroit lobbyists in Iowa history presided over the railroad machine. To permit closer supervision, they divided their areas of operation. N. M. Hubbard of Cedar Rapids, attorney for the Chicago and Northwestern Railway, handled the northern half of Iowa; and J. W. Blythe of Burlington, attorney for the Chicago, Burlington and Quincy Railroad, looked after the southern half. The control that Blythe exercised was so complete that for many years southern Iowa was known as the "Burlington Reservation." Through the extensive distribution of business and transportation favors in the towns and cities of Iowa, the railroads maintained a predominant influence in the Republican party until the death of Blythe in 1909.

Since the railroad leaders who dominated the Republican party could not satisfy all of the demands that were made upon them, they were forced to disappoint several promising political leaders who subsequently became outstanding opponents of the railroads. In the state convention of 1875, General James B. Weaver appeared to have cinched the Republican nomination for governor. At the last moment, however, powerful Republican leaders in the dominant faction became uneasy about Weaver's alleged radical tendencies and his idealism on the prohibition issue. As the roll call vote began, the name of the venerable Republican war governor, Samuel J. Kirkwood, was suddenly thrust into the contest. General Weaver saw almost certain victory elude his grasp as the convention was stampeded for Kirkwood (Cole, 1921:393). Since Weaver realized that his political ambitions were doomed in the Republican party, he began to seek other opportunities for his political skills and new sources of support.

The demise of the Anti-Monopoly-Democratic party coalition in 1875 had left rural interests that opposed the dominant railroad influence in the Republican party virtually without a political voice or leadership. In large measure, this void was filled by a new movement, the Greenback party, which sought to solve the economic ills of Iowa farmers by placing an increased supply of money in circulation. Shortly after his departure from the Republican party, General Weaver became a leading personality and the acknowledged spokesman for the Greenback party in Iowa. Although Greenback support was confined to a relatively small and militant band of farmers, their merger with the Democratic party again was sufficient to produce an impact on Iowa politics. "Fusion with the Democrats sent James B. Weaver to Washington from Iowa, where the Greenbackers polled their highest vote in the country in 1878" (Cole, 1921:66). With his election to Congress, Weaver became the principal leader in both the state and the nation not only in the Greenback party but also in most of the agrarian political movements that were to follow. Despite the support for the Greenback party in Iowa revealed by the election of 1878, however, the subsequent farm movements that Weaver led were no more successful in Iowa than they were in many other states.

When Weaver was himself a candidate for president on the Greenback ticket in 1880 and on the Populist party ticket in 1892, he failed to receive more than 10 percent of the vote in his native state. Furthermore, rural support for both the Greenbackers and the Populists generally did not extend beyond the areas in which farmers had suffered the most severe economic adversities (Fine, 1928:69). Apparently there were still many farmers in prosperous areas that lacked economic or political incentives to join the agrarian movements. The addition of Democratic votes through several fusion efforts greatly enhanced the prospects of Weaver and the agrarian movements. In 1896, for example, when the Populists were included beneath the banner of William Jennings Bryan, they polled a large vote in Iowa. As in most elections during this era, however, the combined strength of the Democrats and the third party did not pose a serious threat to the leaders that controlled the Republicans. The third party movements were an important development in Iowa history because they represented the growing opposition among farmers to railroad influence in the Republican party, but neither the Greenbackers nor the Populists were able to attract a sufficient number of rural voters to substantially alter the course of Iowa politics.

In large measure, however, the agrarian third party movements also probably failed to achieve their goals because they were divorced from the Republican party. Since a substantial number of farmers as well as the residents of small towns and cities remained wedded to Republicanism by memories of the Civil War, the prospects that the third parties could attract the votes needed to produce a major realignment in Iowa politics always were remote. In addition, except for the leadership of General Weaver and occasional mergers with the Democrats, the third parties lacked the organizational machinery necessary to maintain themselves as viable and enduring political associations. By diverting their efforts to sporadic third party movements, the protesting farmers failed to capitalize on the occasion to leave their imprint on the Republican tradition in Iowa politics. Although the agrarian movements indicated that there was a substantial amount of conflict between farmers and business interests, the Republican tradition was still sufficiently strong to prevent the formation of permanent urban-rural cleavages. The Anti-Monopoly, Greenback, and Populist parties lacked both the resources and the continuity to offer a basic challenge to the urban-based railroad machine of the Republican party. For many years, therefore, the railroad interests maintained a powerful influence in the Republican party and the politics of Iowa.

The first major leader within the Republican party to criticize the abuses of railroad dominance was Governor William Larrabee. During his seventeen years in the state Senate and his first term as governor, Larrabee was widely regarded as a "safe" Republican by the railroad interests. As a wealthy mill owner in Clermont, he had obtained substantial public concessions for the railroads to spread the network of rails in northeastern Iowa. His gubernatorial nomination in 1885 had been secured by Colonel David B. Henderson, a former member of the military staff of General Dodge and a leader of his faction in the Republican party (Sage, 1956:72). In 1881 Henderson had written a friend of Allison (quoted in Sage, 1956:181) that Larrabee "is one of the soundest of Republicans and one of the truest of men. . . . All our interests as businessmen and as Republicans will be safe in his hands."

Governor Larrabee, however, accepted his reelection in 1887 as a mandate to accomplish railroad reforms. In his second inaugural address (quoted in Cole, 1921:465-466), he attacked the railroads in proclaiming: "By granting special rates, rebates, drawbacks, and other favors here and there to men of influence in the respective localities, they have secured the favor of many who, after having divided with them their spoils, are ready to defend their wrongs and to advocate a policy of neutrality on the part of the state." Larrabee's challenge aroused the bitter opposition of the railroads. Although he was unable to obtain legislation to outlaw the distribution of free passes or to establish maximum passenger rates, Larrabee did succeed in strengthening the railroad commission. The duties of the commission were changed from strictly advisory to policy-making functions by giving it the power to fix maximum freight rates. Larrabee also appointed men to the commission who supported his aim of curbing railroad abuses. During his second term some reforms were accomplished; but the political retirement of Larrabee in 1889 ended, at least temporarily, the assault on the Republican faction controlled by Blythe and Hubbard.

Along with the dispute over railroad regulation, the Republican party was sharply divided by the question of prohibition. In 1889 the Republican platform unequivocally urged the adoption of legislation to forbid the licensing of the sale of liquor. The Republican position permitted the election of Horace Boies as the only Democratic Governor of Iowa during the latter part of the nineteenth century. While Boies attracted the favor of some rural voters who opposed the influence of the railroad faction in the Republican party, his chief support was drawn from "a great number of Republicans in the river counties, where the sentiment for license was strong" (Haynes, 1916:197). Since most of the votes for Boies were cast by Republicans in the urban areas along the eastern border of Iowa who resisted prohibition, the Democratic party lost the opportunity to champion the rural forces that sought railroad reforms. The Republicans had not resolved their differences over the liquor issue in 1891, and Boise was reelected governor. In 1893, however, the Republicans recaptured the office of governor when the Republican platform (quoted in Clark, 1912:222-223) declared that "prohibition is not a test of Republicanism." Subsequently, a Republican legislature adopted the so-called "Mulct Law" providing for local option on the sale of liquor. Although the Democrats continued to support the relaxation of restrictions on liquor for many years, the question was soon overshadowed by other controversies in Iowa politics.

Within a few years, the opponents of the railroad influence in the Republican party found a new and exciting spokesman in Albert B. Cummins, a prominent Des Moines lawyer. Although Cummins had achieved fame as the attorney for the farmers who broke the barbed wire monopoly in Iowa, he was primarily a corporation lawyer with extensive business connections. In 1894 Cummins decided to become a candidate for the United States Senate. He was totally outmaneuvered in the legislature, however, by the dominant faction of the Republican party which nominated John H. Gear, Governor of Iowa from 1878 to 1882 and the father-in-law of J. W. Blythe. The ambition of Cummins undoubtedly was fired by his defeat; but for the moment, he exchanged his senatorial aspirations for the position of Republican national committeeman.

The next major challenge to the railroad interests was launched in 1897 when state Senator A. B. Funk sought the Republican gubernatorial nomination, despite the opposition of Blythe and Hubbard. The railroad lawyers and "Ret" Clarkson settled upon Leslie Shaw, who had proven his mettle as a popular campaigner in 1896, as their gubernatorial candidate. Although Shaw defeated Funk in the state convention, the contest "marked the definable beginning of the division of the party in the state into what by 1901 came to be called the 'Progressive' and the 'Standpat' factions" because it revealed "with almost positive clarity that the Republican party of Iowa was being 'managed' " (Sayre, 1958:116).

Although the Cummins-Funk faction was to emerge as the spokesmen for the predominantly rural forces that opposed the Blythe-Hubbard wing of the Republican party, their original purpose was devoted to acquiring the public offices that they had been denied. Far from waging a total battle against railroad interests, the evidence indicated the early progressive movement received support from railroad organizations.

> Shortly after the nomination of Shaw in 1897, Cummins wrote Robert Mather, general counsel for the Rock Island Railroad, asking his help in the fight for Gear's seat and suggesting that Mather and his Iowa attorney, Carroll Wright, "occupy toward my contest the same relation that J. W. Blythe occupies toward Gear's." . . . Blythe and the Burlington, aided by Hubbard and the North Western, backed Gear. The Rock Island politicians, Funk and other anti-Blythe Republicans supported Cummins, who had long been attorney for the Chicago & Great Western Railroad in Iowa [Sage, 1956:155].

Cummins initially did not intend to seek the destruction of the railroad influence in the Republican party, but events soon moved him closer to that position.

The frustrated ambitions of the Cummins-Funk faction again reached a breaking point in 1901 when Governor Shaw appointed Congressman Jonathan P. Dolliver to the vacancy in the United States Senate created by the death of Senator Gear. Cummins temporarily postponed his senatorial hopes; and, at a mass rally in Des Moines, he dramatically, announced his candidacy for governor (Cole, 1921:516). In the campaign that followed, Cummins shifted his strategy from a personal appeal to an attack on the railroad machine generally.

> The burden of most of Cummins' speeches was the same—that corporate influence should be driven out of politics and that political power should be returned to the people. Specifically, he talked again and again about the necessity of raising railroad property assessments. . . . Cummins' opponent sought to make the issue that the fight was simply one faction of railroads against another, but his clear attack on railroads in general would seem to break down the argument rather effectively [Sayre, 1958:148-149].

By denouncing the general influence of railroad and business interests in the Republican party, Cummins broadened his support to include the many opponents of the Blythe-Hubbard faction who were concentrated in rural sections of Iowa. Against the new coalition formed by Cummins, the railroad leaders were unable to retain control of the party machinery. At the Republican state convention in 1901, the delegates overwhelmingly nominated Cummins and adopted a platform containing a bitter condemnation of corporate power as well as the famous "Iowa Idea" on tariff reciprocity.

After Cummins had been reelected to a three year term as Governor because of a constitutional amendment changing the date of Iowa elections, in 1906 former Congressman George O. Perkins, a Sioux City publisher, was advanced as the man who could restore harmony to the party. Failing to receive a promise from Perkins that he would appoint Cummins to the United States Senate in the event of Senator Allison's death, however, the Cummins-Funk faction changed its strategy. Cummins re-entered the gubernatorial campaign and defeated Perkins in the state convention for an unprecedented third term as Governor. By 1908, however, Cummins could no longer postpone his senatorial ambitions. Taking advantage of the newly enacted primary law, he announced his candidacy for the Senate seat held by the ailing William B. Allison.

Although Allison was too ill to return to Iowa for the election, the railroads conducted strenuous and extensive efforts in his behalf to thwart the ambition of Cummins. A prominent issue in the campaign was the so-called "Torbert letter" that Cummins allegedly had written in 1906, pledging himself not to be a candidate against Allison (Sage, 1956:221). Although Cummins vigorously denied the letter, his ambitions upset a large segment of Republican voters in Iowa. Not only had he violated tradition by seeking a third term as governor, but he also supposedly had broken a prior promise not to oppose a Senator of extended service and distinction. Despite the intensity of factional issues, many Republicans in both urban and rural areas regarded his lack of respect for the tradition of succession as a major breach of the code of Iowa politics. In the first primary election held in Iowa, Cummins was defeated by the absent Senator Allison for the United States Senate.

Shortly after the primary election, however, Senator Allison died, and his seat was again at stake. Governor Cummins quickly announced his candidacy, and the Standpatters selected former Congressman John F. Lacey as their standard bearer. In another special primary, the Governor finally secured a seat in the United States Senate.

The Republican party faced another serious conflict in 1910 with the death of Senator Jonathan P. Dolliver, who had joined with Cummins to emerge as a prominent progressive in the Senate. Although both the "progressive" and "standpat" factions eagerly contested the vacant position, the standpatters gained an initial advantage when the interim Senate appointment was awarded to Lafayette S. Young, conservative editor of the Des Moines *Capitol.* After a prolonged and bitter fight, however, William S. Kenyon, a mild insurgent from Dolliver's home town who had not been prominently identified with either faction, was elected.

The elevation of Kenyon to the Senate restored a measure of unity to the party. While it was clear that a person of moderately progressive leanings such as Kenyon would be the only candidate acceptable to the Republicans of Iowa at the time, the choice of Kenyon was not a definite victory for either faction. His selection indicated, however, that the standpat faction could no longer exercise the unchallenged political power that had prevailed during the days of the railroad machine. Lafayette Young wrote the epitaph for the standpat faction in Iowa when he spoke editorially in the Des Moines *Capitol* (June 5, 1912), "Standpattism . . . is dead and buried." In large measure, "Kenyon's election sounded the death knell of Iowa factionalism. Fortunately for the Iowa Republican party, it occurred before the disastrous split in 1912" (Visser, 1957:74). Under the general leadership of Senator Cummins, "Iowa was to be a snug harbor for safe progressivism" (Sayre, 1958:420).

The progressive movement probably reached its final flowering in the campaign of 1912. In many ways, however, the national Progressive party led by Theodore Roosevelt was still-born in Iowa. The progressive movement that Cummins led generally was unwilling to follow Roosevelt in part because of prior experience with third party movements. A fundamental reason, however, for the failure of Cummins and his followers to join the Progressive party in 1912 was their strong commitment to the Republican party. The progressive movement in Iowa "was within the framework of the Republican party and strongly within its tradition. Even in their periods of defeat, there seems to be no evidence of a break with the party or its candidates on the part of the progressive leaders" (Sayre, 1958:315-316).

> Cummins found himself in 1912 in the compromising position of announcing his vote for Theodore Roosevelt but refusing to leave the party. Many of his Progressive followers did the same. A few of the more radical deserted to the newly-created but pre-doomed Progressive Party in Iowa. Still others, eager for additional reforms, joined splinter groups, and centered their efforts on such specific issues as constitutional prohibition, women's suffrage, Sunday observance, or the more mundane "good roads" movement. Thus a type of factionalism continued to color Iowa politics for years and the terms "Standpatter" and "Progressive" appear not infrequently; but they had lost their original meaning of denoting membership in distinct political factions which for all practical purposes were distinct political parties [Visser, 1957:77].

The breach between the opposing factions had healed sufficiently by 1912 to prevent the "Bull Moose" movement from disrupting the Republican party of Iowa. The defection of some Roosevelt progressives enabled the Democrats to carry the state for Wilson, but most state and local offices remained in the hands of Republicans who were attached at least to a moderately progressive party.

Although the agrarian third party movements of the late nineteenth century had been supported largely by farmers, the Progressive party led by Theodore Roosevelt apparently attracted a different group of men. A survey (Potts, 1965) of approximately 100 leaders of the Roosevelt, Cummins, and Standpat wings of the Republican party in 1912 revealed that the Roosevelt followers contained about half as many farmers or residents of rural areas as the Cummins or the Standpat factions. On the other hand, only 21.4 percent of the Cummins men as opposed to 37.5 percent of the Roosevelt leaders and 33.1 percent of the Standpatters lived in cities of from 10,000 to 50,000 population. The Roosevelt movement clearly appealed to urban businessmen and professionals, but the differences between the Cummins and the Standpat factions were not as clearly marked. The Cummins leadership was composed primarily of small businessmen and lawyers, while the Standpatters included a large proportion of executives and professionals in "the highest economic bracket." Despite the fact that Cummins had scored some of his initial successes with the aid of rural opponents of the railroads, by 1912 the cleavage between the Cummins and Standpat factions seemed to reflect a division along social and economic lines rather than a distinct conflict between rural and urban interests.

For several years, the Republican party remained relatively undisturbed by the type of intense factionalism that had existed prior to 1912. Ironically, the next major Republican progressive movement was directed against Senator

Cummins and the Esch-Cummins railroad law which seemed to favor business interests at the expense of farmers and workers. In 1920, Cummins was opposed in the Republican primary by Smith Wildman Brookhart, a former Cummins follower and a lawyer from Washington, Iowa.

> Brookhart made his formal announcement on March 20 and accused Cummins of abandoning his old principles and truckling to Wall Street and the railroad executives. He made an appeal to farmers, already pinched by post-war deflation, and laborers, whose wages had never caught up with the cost of living. It was to be the basis of a new crusade. As the years went on, even more than in 1920, Brookhart led a movement of genuine economic protest. He did not attract manufacturers like Maytag, editors like Funk and English, or capitalist farmers like Garst [Sayre, 1958:514-515].

Brookhart's main support was drawn from organized labor and militant farm organizations such as the Farmers Union.

Cummins, on the other hand, relied upon both his old band of small "capitalist progressives" and some new conservative sources. His campaign manager was former Governor and Secretary of the Treasury Leslie M. Shaw, who had once denied Cummins the Senate seat he craved. In the primary election, Cummins also employed the mailing lists of the Iowa Farm Bureau Federation, which was less radical than the Farmers Union, to enlist support. Nonetheless, Brookhart received 45 percent of the primary ballots and "if the vote of the leading city in several counties were omitted, Col. Brookhart would have carried considerably more than half the counties in the state" (Des Moines *News*, February 15, 1922).

Brookhart brought the revived contest between farmers and business interests within the framework of the Republican party rather than allowing it to dissipate in third party movements. Although Brookhart's faction consisted of both rural farmers and urban laborers, his principal strength was concentrated among a new breed of agrarian insurgents who faced growing economic distress during the 1920s. In 1922 Brookhart made his second assault upon the Senate when Senator Kenyon resigned to accept a federal judgeship. Instead of relying upon one major leader, conservative Republicans attempted to throw the election into a special party convention by entering several candidates in the primary to prevent Brookhart from winning the 35 percent of the vote required by law for the primary nomination. Included among the candidates was Clifford M. Thorne, a former state Railroad Commissioner and a lobbyist for the Iowa Meat Producers Association in opposition to the railroads, who shared Brookhart's political sympathies as well as his home town. By 1922, however, the worsening economic conditions had solidified Brookhart's strength in the rural areas of Iowa, and he won the primary with 41.1 percent of the total vote.

Since Kenyon's term would have expired in 1924, Brookhart was faced with the immediate prospect of seeking reelection. Unable to defeat Brookhart in the primary, his opponents sought to discredit him by associating Brookhart with the third party candidacy of Senator Robert La Follette for president. Finally, in a speech on October 3 Brookhart launched a vigorous attack on President Coolidge, although he did not openly advocate the election of Senator La Follette. The Republican state central committee reacted by withholding organizational support from Brookhart's candidacy, and most of the influential Republican newspapers in the state advised their readers to support Brookhart's Democratic opponent, Daniel F. Steck. The newspapers even went "to the extent of publishing specimen ballots, together with detailed instructions as to how a Republican voter could 'scratch' his ballot for Steck. The square opposite Steck's name where the voter was to put a cross was very often indicated by an arrow" (Neprash, 1932:51). After a bitter and hotly fought campaign, Brookhart was declared the winner by a majority of only 740 votes.

Steck contested the election, however, in the United States Senate which is the sole judge of the qualifications of its members.

> He contended that a large number of ballots marked with an arrow pointing to the cross by his name which the Iowa Electoral Commission threw out should have been counted for him since it had plainly been the "intent of the voter" to cast his vote for Steck. If these ballots were included, Steck would be elected. According to the Iowa law, "intent of the voter" is not taken into consideration and these ballots were void. However, the Senate decided that they were valid and declared Steck elected [Neprash, 1932:52].

The 1924 senatorial contest represented one of the few elections in Iowa history in which major leaders of the Republican party publicly encouraged voters to support a Democratic candidate. The results also dramatically illustrated the influence that the press and prominent Republican spokesmen exerted on the voters. For the Democrats, on the other hand, the election exemplified a difficult dilemma. Since Brookhart had kept most of the political strength generated by agrarian unrest within the Republican party, the most effective strategy for the Democrats required that they appeal to more conservative rather than to progressive voters. In 1922, the state chairman of the Democratic party complained (Des Moines *Register*, November 9, 1922) that "the vote for Brookhart was a protest against the Administration and that Brookhart stole Democratic thunder." In serving as a spokesman for the political demands of farmers, Brookhart strengthened the Republican party by demonstrating that it could encompass agrarian protest and by preventing the Democrats from capitalizing upon the sentiments of rural voters. At the same time, the intense factionalism stimulated by the Brookhart movement weakened the Republicans by reinforcing the basis for persistent conflict between rural and business interests and by stimulating some defections to the Democrats.

When Brookhart faced Senator Cummins again in 1926, he received more votes than had ever before been cast for any candidate in a senatorial primary. The Republican party organization was forced to accept Brookhart, and he finally won his first full term in the Senate. Although Henry Field, a folksy Shenandoah nurseryman and radio personality, rallied conservative Republicans to defeat Brookhart in the Republican primary six years later, his movement had a major impact on Iowa politics.

The coalitions formed by Brookhart perhaps reflected the increasing gulf between farms and urban settlements more clearly than any prior movement within the Republican party. In every election in which Brookhart participated after 1920, his vote was negatively correlated with the degree of urbanization of Iowa counties (Neprash, 1932:120). The candidacy of Brookhart represented one of the most dramatic renewals of the conflict between farmers and business or railroad forces that had become a significant feature of Iowa politics.

By 1932, however, the importance of the factional strains between rural and urban interests in the Republican party had been diminished by national political and economic trends. In the Roosevelt landslide of 1932, the Democratic party emerged as a new alternative in Iowa politics. Although farmers were among the first voters to abandon their Republican loyalties, the depression had a significant impact on urban and rural residents alike. The Republican party could no longer afford the luxury of intense factionalism. During the New Deal era Democrats enjoyed an unprecedented taste of political power in Iowa, and Republicans sought to reunite to regain their former position.

Although Democrats were victorious in nearly all elections for state and local offices in Iowa from 1932 through 1936, their success was shaped largely by economic conditions. Once the state had passed through the most severe phases of the depression, the electoral preferences of a majority of Iowa voters abruptly reverted to the Republican party. In their concerted efforts to reclaim political office in Iowa, the Republicans were able to avoid the divisive factional disputes that had racked the party before the depression.

Although the revival of urban-rural conflict occasionally has produced major upsets in Iowa politics, there have been few signs of tension between farmers and business interests in the Republican party since World War II. The durable and organized factions that often clashed over Republican nominations prior to the depression generally have been replaced by loose, amorphous groups that are sometimes induced into primary contests by the personalities of major candidates. In Republican primaries in Iowa, candidates frequently have attracted their largest votes from personal followers in their home counties and surrounding localities. Neither pre-primary party endorsements nor organized slates of opposing candidates in primary elections have been features of recent politics in Iowa. By avoiding the disruptive effects of vigorous factional contests in primary elections, the Republican party frequently has been able to maintain a reduced but persistent margin over the Democrats in Iowa.

On occasion, however, major groups in Iowa politics have been stirred to vigorous competition within the Republican party. In 1948, for example, the Iowa Farm Bureau, which wanted to improve farm roads without disturbing property taxes, became dissatisfied with the policies of incumbent Republican Governor Robert Blue. Along with labor and educational groups, the Bureau supported state Senator William Beardsley who defeated Blue in the gubernatorial primary. The election was unusual not only because it represented the renewed activity of an agricultural organization in a Republican primary but also because the vote did not reflect the support of followers in the area surrounding the candidate's county of residence. Unlike most candidates in modern Republican primaries, Blue failed to carry his home county and Beardsley secured a majority of the vote in most urban and rural counties of the state.

Although the general tendency of the Republican party to avoid factional conflict between rural and business interests since the era of Franklin Roosevelt has strengthened the unity of the party, it also has reflected increasing political competition in Iowa. The depression tended to loosen the traditional Republican attachments of major segments of the Iowa electorate and to elevate the Democratic party as a viable channel for the expression of political dissatisfaction. When economic or political circumstances have operated to the disadvantage of urban or rural residents, voters in both areas have increasingly supported the Democratic party. In the general election of 1948, for example, declining farm prices and urban opposition to Republican labor legislation carried Iowa for President Truman and former Congressman and Senator Guy Gillette. The election returns revealed that "a combination of heavy Democratic voting by labor in the industrial centers and by farmers . . . won the state for Truman" (Des Moines *Register*, November 6, 1948). The same coalition of farmers and laborers returned Gillette to the Senate on Truman's coattails.

Within eight years, the combined effect of urban concern about sales taxes and rural unrest over property taxes in the face of dwindling farm income again seemed to produce circumstances that were ripe for political upheaval. As a national magazine (*Time*, October 22, 1956) observed, "in a state that likes a comfortable status quo as much as Iowa, such a situation–expanding industrialization, squeezed agriculture, uneven economic conditions and higher state taxes–means political trouble for someone." In 1956 that someone was incumbent Governor Leo Hoegh and the Republican party.

Of all the problems that beset Iowa in 1956, the question of taxation was primarily responsible for the defeat of Hoegh and for the election of Herschel Loveless as the first Democratic Governor of Iowa since the New Deal. A survey (Central Surveys, 1956) conducted for the Republican party one month before the election revealed that taxes was the issue "that is hurting Hoegh and helping Loveless." In part the 1956 election represented a negative reaction to Governor Hoegh and Republican tax policies rather than a positive gain for the Democrats. The survey also found that "14 unfavorable references to Loveless compare with a total of 126 unfavorable comments about Hoegh that are given by Loveless supporters." Although Loveless undoubtedly benefited from the early defections of economically disadvantaged farmers to the Democratic party, his support was drawn from voters who opposed a tax increase at nearly all socioeconomic levels and in all geographical areas of the state.

In 1957 Loveless strengthened his popularity by vetoing a one-half percent increase in the sales tax passed by the Republican legislature, and in 1958 he was reelected governor over his Republican opponent who favored a full one percent sales tax rise. At the same election, the cumulative effects of an agriculture recession enabled the Democrats to capture four of Iowa's eight congressional seats and to elect a number of minor state officials.

The Democrats accomplished another major upset in 1962 when Commerce Commissioner Harold Hughes defeated incumbent Republican Governor Norman Erbe. One of the principal issues in the campaign was the advocacy by Hughes of the legal sale of liquor by the drink in Iowa. Since 1933 Iowa law had permitted only the distribution of liquor by the bottle at state owned liquor stores. In May of 1962, however, a newspaper investigation (Des Moines *Register*, May 6, 1962) discovered that over 1,000 federal retail liquor stamps had been purchased by private establishments in a two year period and that liquor by the drink was being sold illegally in two-thirds of the counties of Iowa with little interference by local law enforcement officials. In August a survey (Central Surveys, 1962) reported to the Republicans that the extensive "popularity of liquor by the drink poses a problem. Iowans now say they favor this by more than a 2 to 1 margin. . . . Generally the public identifies the Democrats as the party that favors and the Republicans as the party that opposes liquor by the drink." During the campaign, Hughes gained his greatest support in large towns and cities, while Erbe increased his strength in rural villages and farming areas.

In 1962 Harold Hughes became the fifth Democrat to win a gubernatorial election in Iowa since the Civil War. Commenting on the victory of Hughes, one Republican leader conceded that "if you plot the election returns on a map, you will see a lot of liquor by the drink in the election." The vote revealed that Hughes had carried generally the same urban and "wet" areas of the state that had supported Horace Boies in 1889. In 1963 a Republican legislature, hoping to avoid prior mistakes, passed a bill legalizing the sale of liquor by the drink in Iowa.

The urban vote that Hughes captured, however, may not have been based solely on the liquor question. At the same time, a number of other issues such as reapportionment, taxation, and government reforms were producing significant realignments in the Iowa electorate. In 1964 the popularity of Hughes and extensive dissatisfaction with the Republican presidential nominee permitted the Democrats not only to retain control of the governorship but also to elect their candidates to nearly every major state and local office in both urban and rural areas of Iowa in a landslide that paralleled, if not exceeded, the historic sweep of 1932. In 1966, Hughes was reelected to a third term as governor; and, in 1968, he won a hard-fought campaign for the U.S. Senate. Although the appeal of Hughes to Iowa voters may have played a major role in the revitalization of the minority party, the consistent gains made by the Democrats during the preceding thirty years indicated an increasing possibility of intensified party competition in Iowa.

Since the New Deal era, the Democratic party has acquired growing support among both farmers and the residents of large towns in Iowa. The depression relaxed the party allegiances of many Iowa voters who were formerly wedded to the Republican legacy inspired by the Civil War. In large measure, the persistent Republicanism of the state was stimulated by a political tradition that tended to mute serious conflict; and Republican loyalties were sustained by the ability of the party to absorb the rival demands of farmers and urban residents. Although the aspirations of urban and rural voters have created complicated and troublesome patterns in Iowa politics, in part the problems also concerned the identification of rural-urban conflict. For many years, tensions between farmers and business interests loomed as the major factors in the Republican party and in Iowa politics. Increasingly, however, the Democratic party has tended to replace Republican factional conflicts and sporadic third party movements as the principal vehicle for the expression of urban and rural interests.

REFERENCES

BERGMANN, LEOLA NELSON. (1956) "Immigrants in Iowa." The Palimpsest 37 (March): 135-137.

BLACKHURST, JAMES. (1959) "An examination of the relationship between the Homestead Act and voting behavior in the state of Iowa." Unpublished paper, Syracuse University, Syracuse, New York.

BRAY, THOMAS JAMES. (1957) The Rebirth of Freedom. Indianola: Record and Tribune Press.

CENTRAL SURVEYS, INC. (1956) "State of Iowa, the race for governor, October 8-12, 1956." Unpublished report, Shenandoah, Iowa.

___(1962) "Political opinion survey, state of Iowa, the race for governor, August 13-21, 1962." Unpublished report, Shenandoah, Iowa.

CLARK, DAN E. (1912) History of Senatorial Elections in Iowa. Iowa City: The State Historical Society of Iowa.

___(1917) Samuel Jordan Kirkwood. Iowa City: The State Historical Society of Iowa.

CLARK, OLYNTHUS B. (1911) The Politics of Iowa During the Civil War and Reconstruction. Iowa City: The Clio Press.

COLE, CYRENUS. (1921) A History of the People of Iowa. Cedar Rapids: The Torch Press.

DYKSTRA, ROBERT R., and HARLAN HAHN. (1968) "Northern voters and Negro suffrage: the case of Iowa, 1868." Public Opinion Quarterly 32 (Summer): 202-215.

FINE, NATHAN. (1928) Labor and Farmer Parties in the United States, 1828-1928. New York: Rand School of Social Science.

GATES, PAUL W. (1964) "The homestead law in Iowa." Agricultural History 38 (April): 67-78.

HAYNES, FRED E. (1916) Third Party Movements Since the Civil War with Special Reference to Iowa. Iowa City: The State Historical Society of Iowa.

___(1919) James Baird Weaver. Iowa City: The State Historical Society of Iowa.

KEY, V. O., JR. (1958) Politics, Parties, and Pressure Groups. New York: Thomas Y. Crowell Co.

MILLSAP, KENNETH F. (1950) "The election of 1860 in Iowa." Iowa Journal of History 48 (April): 97-120.

NEPRASH, JERRY ALVIN. (1932) The Brookhart Campaigns in Iowa, 1920-1926. New York: Columbia University Press.

POTTS, E. DANIEL. (1965) "The progressive profile in Iowa." Mid-America 47 (October): 256-268.

ROSS, THOMAS R. (1958) Jonathan Prentiss Dolliver. Iowa City: The State Historical Society of Iowa.

SAGE, LELAND L. (1956) William Boyd Allison. Iowa City: The State Historical Society of Iowa.

SAYRE, RALPH M. (1958) "Albert Baird Cummins and the progressive movement in Iowa." Unpublished Ph. D. dissertation, Columbia University, New York.

TIME MAGAZINE. (1956) "Against the anthills." Time 68 (October 22): 23-25.

VISSER, JOHN E. (1957) "William Lloyd Harding and the Republican party in Iowa, 1906-1920." Unpublished Ph. D. dissertation, State University of Iowa, Iowa City.

WILLSON, MEREDITH. (1962) The Music Man. New York: Pyramid Books.

WOODWARD, C. VANN. (1956) Reunion and Reaction. Garden City: Doubleday & Co.

YOUNGER, EDWARD. (1955) John A. Kasson. Iowa City: The State Historical Society of Iowa.

Chapter **III**

FOUNDATIONS IN THE ELECTORATE

In Iowa the electorate has been divided into three major segments, each with a political life of its own. Large or medium-sized cities, with a sizeable labor force employed in manufacturing or other industries, generally have produced the largest Democratic vote in the state. Small towns, which frequently have been business-oriented, usually have provided substantial Republican majorities. The vote in farm areas, however, often has seemed to fluctuate widely—in part according to the level of farm prices.

Although the voting patterns of the Iowa population usually have seemed to vary along a continuum based on the size of place, the electorate has been marked by several apparent dividing lines. In general, the level of 10,000 population has reflected not only the division between urban areas and small towns but also a relatively clear demarcation in partisan preferences. The separation of voting statistics from small towns and farm townships also has seemed to correspond with significant social and political differences. By analyzing the voting behavior of cities, towns, and farms, the election returns can provide an important indication of the sources of support for the political parties.

The mean Democratic percentage of the vote for President from 1928 to 1964 in cities, towns, and farms is recorded in Figure 1. Although the voting patterns suggest that each of the three areas responded to general trends in the electorate, they also reveal some interesting differences. Perhaps most striking is the vacillating tendency of the farm vote. Prior to 1928, in the presidential elections of 1916, 1920, and 1924, there was a difference of not more than 8 percent in the partisan choices of cities, towns, and farms (Lubell, 1956a:180). The election of 1928, however, seemed to mark a significant realignment in the Iowa electorate as the Democratic vote in farm townships increased by 17 percent. In 1932, the Democratic farm vote continuted to climb to a nearly unprecedented peak; but, as the effects of the depression receded, farmers generally returned to the Republican column. In 1948, a majority of Iowa farmers again supported the Democratic candidate for president, although they subsequently assumed a prominent role in the Eisenhower landslide of 1952. Significantly, in 1956, the Democratic proportion of the vote in farm townships increased more abruptly than in any other section of the state. By 1960, however, the farm vote had returned to an intermediate position between the vote in small towns and the cities.

Unlike the farm townships, the relative position of the vote in towns and cities has tended to remain somewhat constant. In all elections, small towns have been more Republican than the cities. After the initial impact of the depression, which produced a slight majority for Roosevelt in 1932, voters in small towns on the average quickly

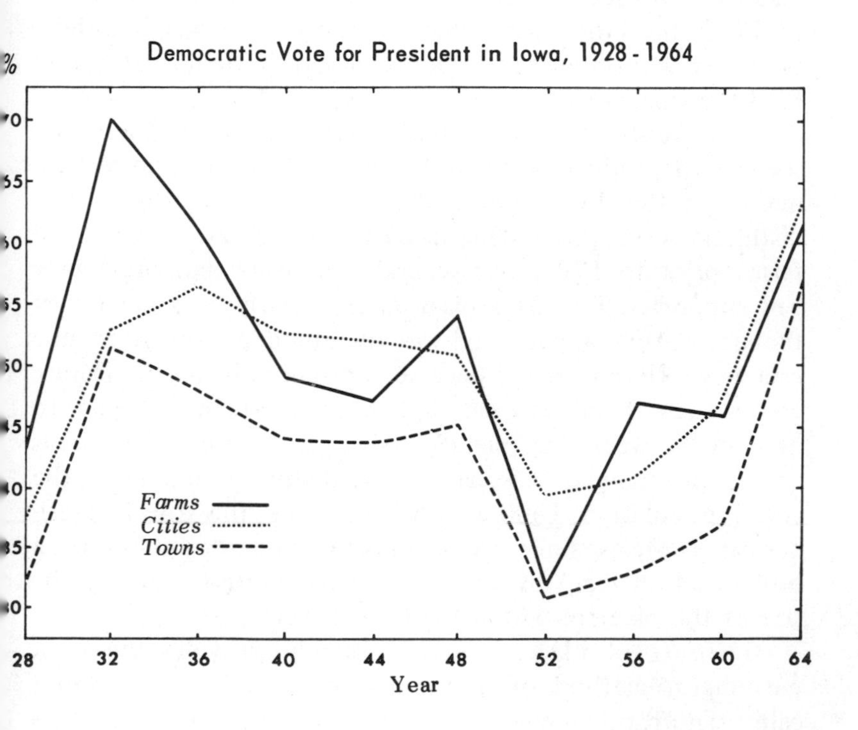
Democratic Vote for President in Iowa, 1928 - 1964
%
Farms
Cities
Towns
28
32
36
40
44
48
52
56
60
64
Year

Figure 1

returned to the Republican fold and remained in that position until 1964. On the other hand, the increase in the average Democratic vote in the cities sustained Democratic majorities through World War II and the election of 1948. In 1952, the cities also joined in the Eisenhower landslide; but they continued to produce a larger Democratic vote than the small towns.

The events that culminated in the depression, therefore, seemed to produce some rather marked and lasting realignments in the Iowa electorate. Although there were few distinctions in the voting behavior of cities, towns, and farms prior to 1928, the general consensus that previously had supported Republicanism among nearly all segments of the population apparently was disrupted by the economic calamity. Urban areas became substantially more Democratic than small towns, and farm townships began to oscillate between the parties. As a result, three major areas or groups emerged as separate and distinct segments of the voting population. Unlike earlier elections, the vote in presidential contests since the depression apparently has been influenced strongly by economic considerations and by the size of the place in which the ballots were cast.

The size of place analysis probably provides the most meaningful method of appraising the tendencies of politically significant segments of the Iowa electorate. Despite the "methodological problem . . . of classifying and then laboriously tabulating elections according to the size of place" (Epstein, 1958:58), the analysis can indicate not only partisan trends among relevant groups in the population but also voting behavior in various categories of communities. The two-party division of the vote by size of place in Iowa in the elections for president, governor, and United States congressman from 1948 to 1964 is contained in Table 1. Possibly the most significant feature of the table is the relatively clear difference in the vote exhibited in nearly all elections at the level of 10,000 population. Cities above 10,000 population generally are more Democratic than towns below 10,000. In addition, the two-party division of the vote in various population categories provides some important insights concerning the results of each election.

Perhaps the greatest controversy has arisen regarding the interpretation of the victory of President Truman in the 1948 election. In Iowa, Democrats garnered a majority of the presidential vote in farm townships and cities above 25,000 population. The vote for Truman in the remaining towns was higher than the Democratic share of the ballots in other presidential elections until 1964, but apparently it was the farm and large city vote that carried the state for Truman.

One analysis (Hardin, 1952:151) of the Iowa vote in the 1948 election, based on county rather than community or township election returns, concluded that "the shift toward the Democrats in counties with less than 25 percent of their income from agriculture was of greater significance in Mr. Truman's victory than the shift toward the Democrats in counties with more than 50 percent of their income from agriculture." Since the plurality for Truman in the 17 latter or urban counties which contributed 41 percent of the state presidential vote was considerably larger than the Democratic margin in the 31 former or rural counties which provided only 21 percent of the total vote, the evidence indicated that the victory for Truman was gained primarily in large cities rather than in farm townships. On the other hand, counties dominated by small towns, which derived from 25 to 50 percent of their income from agriculture and which cast 38 percent of the state vote, gave a majority to Governor Dewey (Hardin, 1955:613).

The county data, however, probably underestimated the

Table 1. SIZE OF PLACE AND THE DEMOCRATIC PERCENTAGE OF THE TWO-PARTY VOTE IN TWENTY IOWA ELECTIONS, 1948-1964

	1948			1950	
	Pres.	Gov.	Cong.	Gov.	Cong.
State	51.1	44.0	45.3	40.7	38.3
Cities:					
Des Moines	55.3	39.6	51.4	44.7	47.0
50,000 and over	57.6	45.2	48.6	44.9	43.2
25,000-50,000	54.4	45.4	51.5	47.2	46.4
10,000-25,000	46.6	38.6	43.0	41.3	35.9
5,000-10,000	45.8	42.7	45.7	35.3	32.6
2,500-5,000	45.6	41.6	40.1	37.0	34.6
1,000-2,500	45.5	41.3	38.6	35.1	32.9
Farm Townships	54.0	46.7	44.6	39.5	36.7

	1958		1960		
	Gov.	Cong.	Pres.	Gov.	Cong.
State	54.1	50.4	43.3	47.9	45.9
Cities:					
Des Moines	60.2	53.8	44.2	50.9	51.8
50,000 and over	59.5	53.2	47.7	52.2	47.8
25,000-50,000	54.4	53.3	46.4	51.9	46.3
10,000-25,000	54.2	48.8	41.6	48.7	44.0
5,000-10,000	46.6	44.8	38.7	41.2	39.7
2,500-5,000	47.2	42.9	37.0	41.3	39.3
1,000-2,500	48.7	45.6	37.4	41.1	39.5
Farm Townships	55.2	53.4	45.6	48.0	48.6

1952			1954		1956		
Pres.	Gov.	Cong.	Gov.	Cong.	Pres.	Gov.	Cong.
35.8	47.9	33.3	48.5	41.5	39.2	51.2	45.6
45.1	54.6	43.6	46.1	46.3	44.1	53.6	50.6
41.3	55.1	50.8	52.5	47.1	39.8	41.8	47.4
43.3	52.4	43.9	57.5	49.6	41.8	52.3	46.6
36.1	50.9	36.0	49.1	41.3	35.8	50.9	44.5
29.4	41.7	31.1	44.0	45.3	31.4	43.0	37.8
31.5	43.8	32.8	42.4	35.6	33.0	44.8	39.1
28.6	41.7	27.2	39.1	33.7	32.6	45.9	39.1
32.0	44.3	32.5	46.8	41.4	47.0	55.0	50.5

1962		1964		
Gov.	Cong.	Pres.	Gov.	Cong.
53.5	46.2	61.7	62.8	51.4
61.6	62.4	68.4	75.8	66.7
58.0	47.6	64.3	73.0	57.2
57.1	49.4	66.6	74.1	59.0
51.0	42.4	61.6	68.2	52.4
46.3	38.8	58.1	60.4	49.1
45.7	39.5	56.4	62.8	46.0
46.0	38.9	58.5	63.6	47.8
49.9	45.4	61.7	61.6	51.3

Democratic majority in farming areas and overrepresented the Truman vote in large cities and small towns. On the basis of community or township voting returns, the Democratic vote in farm townships was four percent higher than in the counties sustained mainly by agriculture; but it was four percent and one percent lower in towns and cities, respectively, than in the equivalent counties dominated by small towns and urban areas. In reality, the unexpected successes scored by the Democrats in Iowa in 1948 probably were produced by a combination of agrarian unrest and urban support. While farmers were concerned about the decline in agricultural prices immediately before the election, city voters were mobilized by the controversy concerning the Taft-Hartley Act and other issues. Meanwhile, however, the vote in small towns remained substantially Republican. By separating the vote in farm townships and small rural communities as well as urban centers, the prominent role of both farmers and city voters in the Truman victory was readily apparent.

In 1950, the vote in all population categories returned quickly to the Republican column. Apparently the relatively volatile farm vote joined other segments of the electorate in an extensive mid-term reaction against the Truman administration. An interesting feature of the election, however, was the campaign for the United States Senate. Albert J. Loveland, Undersecretary of Agriculture in the Democratic administration, resigned his position to campaign for the Senate on a platform promoting the high price support program of Secretary of Agriculture Charles Brannan. Loveland's opponent, incumbent Republican Senator Bourke B. Hickenlooper, opposed the program. Loveland lost the election, but he received 46.2 percent of the vote in farm townships. His Democratic gubernatorial runningmate, Lester Gillette, however, obtained only 39.5 percent of the farm vote. Although Republican strength in 1950 precluded a Democratic upset, the difference between the vote for Loveland and Gillette in farm townships seemed to indicate support for high price support programs among Iowa farmers.

In 1952, the enormous popularity of General Eisenhower among all groups of voters in Iowa produced a Republican victory that extended to nearly all campaigns in the state. The triumph of the traditionally dominant party in that year even exceeded the Democratic landslides of 1932 and 1964. The only major Democratic candidate who appeared as a serious contender in 1952 was the gubernatorial nominee, Herschel C. Loveless, a former urban mayor of Ottumwa. Although a state survey revealed that only 3.5 percent of the voters expressed a particular interest in his candidacy, Loveless ran well ahead of the Democratic ticket in the state. His strength, however, was largely confined to cities above 10,000 population, the only communities in which he received a majority of the vote. Aside from the interest that the Loveless candidacy generated in urban areas, Republicans scored a nearly total sweep of elective offices in Iowa.

The 1954 election, however, revealed potential dissatisfaction with the Republican party at least among some sectors of the electorate. The adoption of the flexible price support program, advocated by Secretary of Agriculture Ezra Taft Benson, apparently inspired some farm families to switch their political allegiances to the Democratic party. In congressional elections, for example, the Democratic proportion of the farm vote increased by 8.9 percent between 1952 and 1954. In the sixth congressional district, the vote in farm townships for the incumbent Republican Congressman, James I. Dolliver, dropped from 70.2 percent to 57.3 percent. Furthermore, in the 1954 race for the United States Senate, the incumbent Democrat, Guy Gillette, who had been returned to the Senate on the coattails of President Truman in 1948, received 52.9 percent of the farm vote but only 49 percent of the total vote in the state (*U.S. News and World Report,* January 7, 1955:37). A vigorous campaign by Democrat Clyde E. Herring, Jr., son of the governor during the depression years, also made the 1954 gubernatorial campaign one of "the closest since 1936" (May, 1955:251). The voting patterns in farm townships seemed to portend the Democratic victories that were to follow in 1956 and 1958. Although the Democratic percentage of the farm vote was not sufficiently large to contribute decisively to the total state vote, there were indications that farm families might join with other groups in the population to produce Democratic majorities in the coming elections.

Despite the renewed effects of the Eisenhower candidacy, the vote in farming areas and large cities was sufficient by 1956 to make Herschel Loveless the first Democratic Governor of Iowa in 20 years. While Loveless acquired substantial majorities in most of the urban areas of the state, he also attracted sizeable support in farm townships. In the sixth congressional district, the incumbent Republican Dolliver also was defeated by Democrat Merwin Coad. In 1956 Coad gained 57.3 percent of the vote in farm townships, the same percentage that his opponent received in 1954. In all congressional districts, the Democratic percentage of the farm vote for Congress increased 9.1 percent between 1954 and 1956. Similarly, in the presidential election, farm townships produced a larger Democratic vote in 1956 than even the major urban centers. While the large cities played a prominent role in the election of a Democratic governor in 1956, the vote that was shifting most rapidly to the Democratic column was cast in farm townships.

In the election of 1958, the Democratic party enjoyed one of its most complete victories in Iowa since the depression. In addition to reelecting Loveless, the Democrats were successful in electing four of eight congressional candidates. Although the results of the balloting commonly were interpreted as indications of a widespread "farm revolt," Democratic gains in the urban areas and some slight increases in small towns seemed to play a more prominent role in the outcome than the farm vote.

In the congressional elections, the vote in farm townships "was slightly more Democratic than the total vote" in seven of the eight congressional districts (*Wallaces' Farmer,* January 17, 1959:26). Among the four districts in which the Republicans were defeated, however, there were only two constituencies in which the Democratic proportion of the vote increased substantially from 1956 to 1958. Furthermore, only in the sixth congressional district was the growth of the Democratic vote between 1956 and 1958 greater in the farm townships than in the major city of the region. In the sixth district, where Democratic Congressman Coad was elected in 1956 and reelected in 1958, the Democratic proportion of the farm vote grew by 8.3 percent, while it increased from 51.4 percent to 56.4 percent in Fort Dodge, the largest city in the district. In the second district, where the rural vote for the Democratic candidate also gained appreciably between 1956 and 1958, the Democratic proportion of the farm vote increased from 46.3 percent to 53.5 percent; but it jumped from 56.8 percent to 63.2 percent in Dubuque, the principal city. In the two remaining districts the evidence was even more conclusive. The Democratic share of the farm vote in the fifth district in 1958, for example, inched only one-half of one percent above its 1956 level of 53.0 percent, while it grew from 50.6 percent to 53.8 percent in Des Moines. Similarly, the Democratic percentage of the farm vote in 1958 in the fourth district declined by one percent from its 1956 mark of 54.0 percent, but it increased from 55.6 percent to 60.7

percent in the major city, Ottumwa. Since the total urban vote also was much larger than the farm vote in nearly all of the districts, the ballots from farm townships probably were not decisive in electing any of the new Democratic Congressmen from Iowa in 1958. Even in the sixth district, the growth of the Democratic vote in farm townships between 1956 and 1958 was overshadowed by the Democratic gain of 14.6 percent in the same areas from 1954 to 1956.

In addition, the gubernatorial election revealed a similar pattern. While support for Loveless increased by approximately five percent in the largest cities of the state and small towns became slightly more Democratic, his proportion of the vote in farm townships remained relatively stationary between 1956 and 1958.

The Democratic victories in the 1958 election, therefore, were not solely the products of a momentary partisan shift among Iowa farmers. In fact, by 1958 the farm vote in Iowa apparently had ceased to play a prominent role in the election of Democratic nominees. As the most volatile segment of the electorate, farmers seemed to respond more quickly to economic adversities and other events than other voters. Democratic tendencies in farm townships were evident in 1954 and 1956; but they had subsided somewhat by 1958, when the Democratic vote increased substantially among voters elsewhere. In large measure, however, the farm vote seemed to mark out a pattern of defection from the Republican party that gradually was followed by other groups. The trend toward the Democrats that was established in farm townships by 1954 and 1956 soon spread to other portions of the state. Perhaps the voting patterns may have reflected the delayed ramifications of a recession in the agricultural economy that subsequently affected increasingly larger groups in the population. Although Democratic gains in farming areas generally were not sufficient to upset Republican officials, they did seem to launch a partisan divergence that eventually extended to the large cities and, in limited measure, even to small towns.

In 1958 and in subsequent elections, other segments of the voting population seemed to contribute more to a departure from traditional Republicanism than farm townships. Unlike some earlier contests, the elections from 1958 to 1964 were marked by a relatively narrow two-party division of the total vote and by a growing tendency for the partisan differences between urban and rural communities to reflect Democratic majorities in the large cities.

In 1960, Iowa seemed to return to the Republican fold as two of the Democratic Congressmen were defeated and the state cast its ballots for Vice-President Richard M. Nixon as well as Republican candidates for governor and other major offices. Although none of the population categories yielded a majority for President Kennedy and only Des Moines cast a Democratic plurality for congress in 1960 or in 1962, Democratic gubernatorial nominees obtained a plurality in most of the cities above 10,000 population. The strongest sources of Democratic support were in the urban areas, while farm townships occupied a partisan position between the towns and cities, and rural communities remained steadfastly Republican.

The strength of Democratic gubernatorial candidates in urban centers probably reflected the growing importance of issues such as taxation, liquor control, and reapportionment that divided urban and rural voters. As the only major officials in Iowa elected by the state electorate in which the urban vote played a prominent role, candidates for governor were in a unique position to espouse urban programs that had encountered rural opposition in the legislature. In 1962, Harold Hughes became the second Democratic governor of Iowa since the depression by campaigning on a platform advocating the legal sale of liquor by the drink. Although Hughes received his primary support in large cities above 10,000 population, he failed to obtain a plurality in farm townships only by the narrowest possible margin. Urban centers provided the principal bases for expanding Democratic majorities in the state, but the farm vote had not become aligned with the relatively consistent Republicanism of small towns in rural areas.

In 1964, state Democrats attained a success in Iowa that even surpassed their victories in the election of 1932. In addition to the support given President Johnson, Democrats succeeded in electing all major state officials, a majority in both houses of the legislature, and six of seven Iowa Congressmen. Although the results of the election frequently were interpreted as a repudiation of Republican presidental nominee Barry Goldwater, the major coattails for local Democratic candidates probably belonged to Governor Hughes rather than President Johnson. In nearly all areas of the state, Hughes led the Democratic ticket by a substantial margin. The primary sources of Democratic votes again were in the large cities, but the small towns also cast unprecedented majorities for Hughes and Johnson. Significantly, however, farmers were more favorable to the Democrats in 1964 than the residents of small towns, and their ballots closely paralleled the two-party division of the vote in the state. While subsequent elections indicated that the Democratic upsurge in 1964 probably was a temporary or peculiar phenomenon, the voting patterns of cities, towns, and farming areas during the period from 1948 to 1964 suggested a set of alignments that may have an enduring impact on Iowa politics.

If the trends of the 16-year period after 1948 eventually produce increased party competition in Iowa, Democrats may have to rely primarily upon the electoral strength of cities above 10,000 population to acquire sufficient votes to offset the traditional superiority of the Republican party. The voting statistics from the nine elections indicate, however, that while urban-rural differences may constitute the principal foundations for renewed party competitiveness, the farm vote frequently may intervene to prevent partisan contests from being divided solely by urban and rural characteristics. In several elections, the margin of victory for Democrats in Iowa seemingly has been provided by a combination of urban and farm ballots. The partisan variability of the vote in farm townships, therefore, probably will continue to impede the development of campaigns into exclusive battles between urban and rural areas.

The size of place analysis of voting in Iowa clearly has tended to refute the notion that the farm vote usually has been overwhelmingly Republican. In large measure, the common characterization of farmers as Republican probably has developed from the confusion of the farm vote with election returns from small towns. By separating the vote from farm townships and rural communities, however, it has become apparent that farmers have not subscribed to the relatively consistent Republicanism of small towns in Iowa. In nearly all elections since 1928, farm residents have been less Republican than the inhabitants of rural towns. The incipient Democratic leanings of farm families also may be cited in interpreting the propensity of the farm vote to reflect partisan trends in the large cities. Since farmers have been more Democratic than small town voters even in Republican years, they frequently have reflected an intermediate position between the vote in urban centers and rural communities.

Perhaps the most striking feature of the farm vote, however, has been its mercurial pattern in relation to the political parties. The ease and speed with which farmers depart from predominant party affiliations also has been observed at the federal level. "In the national elections since 1948, the two-party vote division among farmers outside the South has fluctuated more sharply than it has within any major occupational groupings, or for that matter, within any of the groups considered as voting blocs" (Campbell et al., 1960:402). The experience in Iowa has indicated that farmers have ranked as the segment of the electorate most likely to switch partisan loyalties in state as well as presidential elections. In general, the vacillations in the farm vote have not discriminated between party candidates for various offices.

The tendency of farm voters to differentiate between the parties rather than between elective offices apparently has extended even to candidates who have taken opposing positions on agricultural issues. One study, for example, compared the vote for congressional candidates who disagreed on the farm policies of Secretary of Agriculture Benson in rural districts during the Eisenhower administration. The results (Gilpatrick, 1959:320) indicated "that there was no discernible difference in the way farmers on adjacent farms and engaged in the same type of farming but on opposite sides of a congressional district line voted for Congressmen from the same party when these men had gone on public record on opposite sides of the price support issue." Although his advocacy of a high price support plan may have been of some value to Undersecretary Loveland in the 1950 senatorial campaign, criticism of the Benson program apparently was not sufficient for some Republicans to retain their popularity among Iowa farmers. As the study (Gilpatrick, 1959:323) noted:

> Republican Congressmen . . . generally lost ground with the farmers in 1954 in complete disregard to their individual positions, and they continued to lose ground, although at a lesser rate, in 1956 when one Iowa district went Democrat. By 1958, however, the rate of Republican disaffection had leveled off to an average of only two percentage points per district in Iowa even though three more Republicans were replaced by Democrats.

The initial political reaction of farmers to the policies of the Eisenhower administration seemed to be more severe than their subsequent electoral responses. The sharpest decline in the Republican vote in farm townships occurred in 1954 and 1956 when flexible price support policies and the threat of falling farm prices first appeared.

In general, fluctuations in the farm vote have coincided with decreasing farm prices and low farm income. Since the economic decline of the 1950s did not approach the calamity of 1932, the farm vote was not as overwhelming in its response to economic circumstances. Yet, variations in partisan preferences among farmers generally have been most acute at the first sign of declining agricultural prosperity. As a result, there probably has been an element of truth in the old aphorism that "the Democratic vote in Iowa is determined by the price of hogs on election day." Since the depression, the Democratic party apparently has replaced the third party movements and progressive or insurgent revolts of an earlier era as the appropriate channel for the expression of agrarian discontent in Iowa.

The proclivity of farmers to respond immediately to economic recessions probably has been a larger determinant in their voting than the type of farming in which they are engaged. A study (Neprash, 1932:120) which concluded that farm support for Senator Smith W. Brookhart during the 1930s was "predominantly economically motivated" also found that crop areas were a secondary influence that tended to sharpen rather than modify the political reaction of farmers to economic circumstances and that "enough voters were swayed by economic considerations to determine the outcome of each election." Farmers have differed greatly in the size and type of farming in which they engaged, but such distinctions have seemed to reinforce rather than to alter the inclination of farmers to switch political affiliations temporarily in the face of dropping farm prices. Consequently, economic forces probably have played a greater role than any other factor in the wide fluctuations in the farm vote in Iowa.

The tendency of the vote in farm townships to change its partisan character rapidly also has been indicated by the relatively large number of political independents among farm families. Perhaps because of their peculiar vulnerability to economic pressures, farmers apparently have been less committed to the Republican tradition in Iowa than other groups of voters. In 1958 the *Wallaces' Farmer* (December 6, 1958:24-25) Poll attempted to determine the proportion of Iowa farm families who were closely identified with either the Democratic or the Republican party. Using the criterion of straight-ticket voting in the 1956 and 1958 elections, the survey discovered that sixty percent of the farm voters in Iowa formed the "hard core" of Democratic or Republican party strength in rural farm townships. (Republicans, of course, outnumbered Democrats on this basis.) "The other forty percent were the shifters, the [people] who scratch tickets and who decide elections." A similar division of party adherents seems to have existed in other eras. In the Brookhart elections (Neprash, 1932:120-121), for example, it was found that "though to a great extent the attitudes expressed in these elections were traditional or otherwise determined, still, the attitudes of a sufficient number of 'marginal voters' were economically determined to materially affect the result." The sharp and rapid oscillations in the farm vote probably have been accentuated by the relative absence of persons with firm partisan attachments who might otherwise maintain the vote of a party that is disadvantaged temporarily by its position on an issue to which voters are economically sensitive.

In addition, farm voters often have been less likely to participate in party organizations or other groups than the residents of communities. Since group associations normally have seemed to play a significant role in the formation of political attitudes, the relative lack of group relationships probably has been a contributing factor in the development of the political independence of farm families. In fact, one study (Campbell et al., 1960:428) has concluded that "the lack of firm bonds between the farmer and traditional political group structure is but a special case of a broader insulation of the farmer from group activity." The organizational isolation of the farmer frequently has been reflected in the sporadic and unstable character of agrarian protest movements and in his repeated inability to organize for enduring collective political action. Since farm families usually have evidenced less involvement in continuing political organizations than other voters, they have been subjected to relatively few group reinforcements that often inspire stable partisan allegiances.

The partisan activity of farmers also has seemed to increase as political events assume a growing relevance to their personal fortunes. Since most farm policy questions have been debated on a national level, farmers apparently have demonstrated greater interest in national than in state elections. In 1946, for example, an Iowa Poll (October 21, 1946) discovered that farm families were less likely to vote in all elections but more likely to vote in presidential elections than the inhabitants of towns or cities. In addition, fewer farmers reported that they had never voted than city or town residents. The evidence has suggested (Campbell et al., 1960:405) that "in addition to fluctuations in the rural partisan vote division, there appear to be shifts as well in rural turnout that are wider than is customarily found in more urban districts."

Nonetheless, there have been increasing indications that the farm vote no longer can be considered as a homogeneous political bloc. As agricultural policies have become less controversial at least in some sectors of society, a significant division in the vote has been emerging within farm townships. A *Wallaces' Farmer* (November 19, 1960:17) Poll in Iowa, for example, discovered that, among farm families, there was an important difference between partisan preferences and the age and sex of the voter. In the presidential election of 1960, Kennedy received the support of 61 percent of the farm men who were from 21 to 34 years of age, 53 percent of the men in the 34 to 49 age bracket, and only 41 percent of those who were 50 and over. Among farm women, on the other hand, Nixon was favored by 56 percent of those between 21 and 34, 65 percent of those from 34 to 49 years of age, and 74 percent of those who were 50 or older. Perhaps the association between age and support for the Republican candidates by both men and women was due to the increased influence of traditional Republicanism among older voters. But the Republican sympathies of farm women seemed to suggest the truth of the old adage that "the farmer votes Democratic to raise farm income, while his wife votes Republican to keep her sons out of war." Although farmers have been perhaps particularly sensitive to the economic cycle of agricultural prices, other issues also have affected the partisan preferences of farm women and families.

A *Wallaces' Farmer* (September 20, 1958:52-53) Poll revealed that farm policy and international affairs were the most important issues to farm voters who were not closely affiliated with either political party. The differential impact of the two issues, however, also was evident among farm residents who were firm Republicans or Democrats. In responding to a question about which of the two parties would do a "better job of preventing war," 38 percent of all farm voters cited the Republicans, 13 percent mentioned the Democrats, 28 percent felt that there was no difference between the parties on this issue, and 9 percent were undecided. Significantly, the results reflected a sizeable loss for the Republicans from 1956 when 53 percent reported that they believed that the Republicans would do a better job of preventing war and only 6 percent expressed similar confidence in the Democrats. Even in 1958, however, 72 percent of the farm Republicans stated that their party was more capable of maintaining peace, while none of the Republicans referred to the Democrats in this connection, and the remainder saw no difference between parties or were undecided. Among farm Democrats, on the other hand, 32 percent were willing to confess that the Republicans would do a better job of preventing war, while only 28 percent evaluated their own party most highly and an identical proportion perceived no difference between the parties, with the remaining 12 percent expressing indecision. Despite the trend against the Republicans, therefore, they were judged as more capable of preserving peace not only by all farm residents and by members of their own party but by farm Democrats as well.

Some significantly different patterns, however, were revealed by a similar question asking which of the two parties would do a "better job of raising farm income." The replies indicated that farm voters generally have favored the Democrats on this issue as 59 percent of all farm residents in 1958 and 52 percent in 1956 voiced greater confidence in the Democrats. Only 22 percent of the respondents in 1956 and 11 percent in 1958 ranked the Republicans more highly on farm policy. The advantage of their party on the farm issue was recognized by the Democrats, 84 percent of whom rated their party as better able to raise farm income, while none evaluated the Republicans in this way, and the remainder were either undecided or saw no difference. Among farm Republicans, however, only four percent admitted the Democrats would do a better job on agricultural policy, 49 percent claimed that their party was more capable, 34 percent argued that there was no difference, and 13 percent were undecided. On the farm issue, which seemed to be one of the principal assets of the Democrats among farm residents, there was less willingness by Republicans to grant a concession to the Democrats than there had been in opposite circumstances on the preceding question of preventing war.

The results of the two questions raised some interesting inferences concerning the issues that may have affected voting behavior in farm townships. While Democrats were considered as more capable of raising farm income, Republicans enjoyed a substantial advantage on the issue of maintaining peace. The differing economic relevance of agricultural policies at various times probably has been related to the fluctuations in the farm vote. When economic conditions have made farm programs most salient to farm residents, the vote has tended to gravitate toward the party that they perceived as superior on this issue. When other issues and problems have intervened or assumed greater importance, however, the farm vote has reflected the advantage of the Republicans. Significantly, Democrats were willing to make more concessions to the Republican party on foreign policy than Republicans were inclined to grant the Democrats on the farm problem. In part, this finding may have implied the pervasive influence of the traditionally Republican environment in Iowa; but it also probably reflected the instability and minority status of the Democrats. In normal economic periods, the acknowledged superiority of the Republicans on questions that were unrelated to farm income clearly has had a major impact on voting intentions, even among farmers who were disposed to cast Democratic ballots when agricultural prosperity declined.

In addition, there have been growing indications that the importance of agricultural programs in the partisan preferences of farmers may be overshadowed by other considerations. In part, this trend probably has been stimulated by the "changing work experience of farm men and women" (Murphy, 1960:28). The tendency of farmers to supplement their incomes with non-farm employment has developed to the extent that one Democratic party chairman in Iowa even estiamted that 70 percent of the farm families in his county derived at least a partial source of their incomes off the farm. The sharp decline in the farm population as well as increasingly diversified sources of income probably has reduced the dependence of farm residents on government programs in agriculture and the fluctuations of the market place.

Although the declining importance of farm policy would seem to indicate that the farm vote might return to its traditional Republican moorings, the trend does not indicate that Democratic campaign efforts in farming areas would be totally futile. Young farm voters are more likely to be Democrats than older persons on farms. In addition, as more farm families are employed in manufacturing and other industries, the eventual identification of the farmer may be with urban labor rather than with small town business. While speculation concerning agricultural trends remains difficult, the evidence available thus far suggests that

> farm politics are tending to become urbanized. Farmers generally are being split into two economic classes, each of which appears to be moving toward political alliance with similarly situated urban elements. While the more well-to-do farmers are aligning themselves with the middle-class elements in the cities and suburbs, the less well-off farmers seem to be aligning with the Democratic income elements in the cities [Lubell, 1956b:174].

Although the farm vote gradually may lose some of its distinctively volatile and cohesive features, it probably will continue to occupy an intermediate position between the electoral patterns of urban areas and rural communities.

Since small towns perhaps have been most directly affected by the economic well-being of farmers, the expectation naturally might have arisen that those communities would tend to mirror the oscillations of the farm vote. In general, however, the economic dependence of small towns upon surrounding farm areas has seemed to have little effect on the political behavior of small towns. Towns below 10,000 population have been the most consistent sources of Republican strength in Iowa. The Democratic percentage of the vote in small communities has decreased more consistently as the size of towns declined than it has increased in large cities. Unlike the urban areas and farm townships, small towns in Iowa have provided Democratic majorities in only two elections since the depression.

Not only have small towns failed to produce the Democratic vote that farm areas occasionally have yielded, but they also have evidenced a delayed and rather undramatic response to the voting trends of farm residents. In general, small towns have been more responsive to a decline in the Democratic proportion of the farm vote than to an increase. In 1950 and 1952, rural communities seemed quick to mirror Republican gains in farm townships. After 1952, however, when the Democratic share of the two-party vote began to climb rapidly, there was a noticeable lag in the trend in the towns. Although the usual connotations of the rural category have included both small towns and farm townships, there have been important differences in the voting patterns of the two areas.

In large measure, the Republican proclivities of small towns probably developed from their traditional business orientation. The establishment of most towns usually required the solution of "problems centering on the advancement of business" (Elkins and McKitrick, 1954:342). The towns became commercial centers for farms on the surrounding countryside. "For the townspeople, many of whom were agents of railroads, banks, grain and feed companies, the chain of identification was not with the farmer but with business" (Lubell, 1956a:182). Since the everyday concerns of town residents revolved about business rather than agricultural or labor matters, they gradually developed an affinity with the Republican party.

Most small towns, however, were concerned not only with sustaining a business or commercial center, but they also were zealous in their efforts to expand and improve the economic advantages of the community. Nearly all inhabitants were incorporated in a legacy of "boosterism" that required a firm belief in the eventual prosperity and greatness of the municipality. A "militant, traditional faith in the town's future constituted a hegemony that prevailed in all aspects of community life. In politics, for example, the natural identification with business interests and the demands of this optimistic faith have produced an unshakeable Republicanism in small towns" (Hahn, 1965).

The business-oriented and largely Republican influences of small towns were partially extended to their clients in the country. As the towns were becoming "the natural focus of exchange of goods and services in the Northwest, they must thus inexorably be the focus of politics as well" (Elkins and McKitrick, 1954:342). In addition to their relative social and cultural advantages, many small towns acquired political prestige and importance by serving as the location of county courthouses. Unlike farm townships, small communities possessed several attributes that made them the natural objects of political attention and leadership in rural areas. "Most of the conditions of farm life tended to keep the farmer in a status inferior to the town" (Lubell, 1956a:182). In fact, the historical Republicanism of Iowa probably spread from the towns to the farms rather than in the reverse direction.

In addition, numerous factors in the community life of

small towns have been favorable to the maintenance of traditional political loyalties. The social structure of rural towns in Iowa probably has been more stable than any other similar group in American society. "The village serves as a place of residence for disproportionately large numbers of some of the most dependent groups in American society, and particularly for aged persons of both sexes and for widowed and divorced females" (Smith, 1942:21). Such communities "seldom have the economic advantages to attract residents representing new and different socio-political attitudes" (Epstein, 1958:70).

The social and economic characteristics that have perpetuated traditional party attachments in small towns usually have been reinforced by a climate of opinion that has attempted to prevent political activity and leadership from escaping traditional partisan channels. "In many small cities and villages . . . Republicanism has remained the only approved vehicle for political action. Thus to become a Democrat is not only to join a minority; it is to become a social deviant" (Epstein, 1958:68). For many years, particularly in the small towns of Iowa, a moral opprobrium has been attached to membership in the Democratic party. In reminiscing about his boyhood in Iowa, for example, former President Herbert Hoover (1951:9) observed:

> There was only one Democrat in the village. He occasionally fell under the influence of liquor; therefore, in the opinion of the village, he represented all the forces of evil. . . . [He] served well and efficiently for a moral and political example.

In many small communities, prevailing social sanctions have been effective in perpetuating the Republican tradition.

The Republican consensus in rural towns also has been solidified by the interactions of members of the community. Since most inhabitants of small towns have been familiar with the attitudes and affiliations of other residents, traditional political values usually have been pervasive and dominant. As a result, Democratic candidates commonly have received less support from prominent community leaders than Republicans. In general, partisan activity in small towns has been "predominantly a one-party politics which is reflective of a highly integrated community life with a powerful capacity to induce conformity" (Key, 1956:227).

Unlike the rural communities, which have produced the largest and most consistent Republican vote in the state, the election returns from urban cities of more than 10,000 population normally have been more Democratic than the state average. In the main, an increase in the size of the community has seemed to correspond with an increase in the Democratic proportion of the vote. Although relatively large cities in Iowa have not been as solidly Democratic as heavily populated metropolises in other states, they generally have been the principal sources of minority party strength in the state since the depression.

The sharpest drop in the Republican share of the vote in incorporated communities generally has occurred at the level of 10,000, and, to a lesser extent, at the level of 25,000 population. There have been few appreciable differences in the voting patterns of cities at the top of the population scale. The most populous city in the state, Des Moines, which is more than twice as large as its nearest competitor, typically has not yielded the impressive Democratic percentages that might have been anticipated if the size of place bore an exact relationship to the two-party division of the vote. A similar study in Wisconsin (Epstein, 1958:61) found the most dramatic distinction in the vote separated large- and medium-sized cities approximately at the level of 50,000 population rather than at "the line between a great metropolis and all other cities." While the voting patterns in Iowa generally failed to betray sizeable differences between major and medium-sized cities at the level of 50,000, they did reveal a fairly consistent distinction of about five percentage points between urban and rural communities.

While voting behavior in cities clearly has been affected by a large number of social and economic factors, certain important political characteristics have been highly associated with the size of communities in which the ballots were cast.

> For example, in big cities it may be more likely that workers would live in a distinct neighborhood community, that factories would be larger and more effectively organized, and that more trained, capable, and politically conscious labor-Democratic leadership would be available. Such conditions, clearly associated with size of place, could well be decisive in obtaining Democratic votes in communities, which . . . have been Republican historically. Where these conditions are absent, as they are even more obviously in the smaller than in the medium-sized cities analyzed here, Democratic electoral efforts suffer a relative handicap [Epstein, 1958:66].

The process of urbanization, apart from the influence of other related factors, therefore, can have a significant impact on the two-party division of the vote.

Although the urban characteristics associated with sizeable Democratic pluralities generally have been absent in Iowa cities, there have been growing indications that common interests might solidify the city vote. Since 1956, particularly in gubernatorial elections, urban areas in Iowa have cast a surprisingly heavy Democratic vote. As issues producing urban-rural differences became increasingly clear and pressing, the large cities began to emerge as the principal sources of Democratic votes and as major bases for increased party competition in the state.

While the city vote in Iowa has evidenced some Democratic propensities, it has not approached the unity or the magnitude of the vote in major cities or states. Since urban areas in Iowa have consisted of medium-sized cities rather than extensive metropolitan regions, they may never yield the impressive Democratic majorities that have arisen elsewhere. Yet, Iowa has seemed to develop some relatively firm foundations for intensified partisan competition. Although small towns have evidenced relatively steadfast Republican leanings, large cities have reflected a growing Democratic trend and the farm vote has demonstrated a volatile political independence that could swing an otherwise evenly divided electorate toward either party. As the partisan distinctions between cities and rural communities gradually influence other aspects of Iowa government and politics, urban-rural differences might produce a significant realignment within the state.

REFERENCES

CAMPBELL, ANGUS, PHILIP E. CONVERSE, WARREN E. MILLER, and DONALD E. STOKES. (1960) The American Voter. New York: John Wiley & Co.

ELKINS, STANLEY, and ERICK McKITRICK. (1954) "A meaning for Turner's frontier: democracy in the old Northwest." Political Science Quarterly 49 (September): 321-353.

EPSTEIN, LEON D. (1958) Politics in Wisconsin. Madison: University of Wisconsin Press.

GILPATRICK, THOMAS V. (1959) "Price support policy and the Midwest farm vote." Midwest Journal of Political Science 3 (November): 319-335.

HAHN, HARLAN. (1965) "The lost history of boomtown: some interpretations from Hamlin Garland." Annals of Iowa 37 (Spring): 598-610.

HARDIN, CHARLES M. (1952) The Politics of Agriculture. Glencoe: The Free Press.

——(1955) "Farm price policy and the farm vote." Journal of Farm Economics 37 (November): 601-624.

HOOVER, HERBERT. (1951) The Memoirs of Herbert Hoover: Years of Adventure, 1874-1920. vol. 1. New York: The Macmillan Co.

KEY, V. O., JR. (1956) American State Politics. New York: Alfred A. Knopf.

LUBELL, SAMUEL. (1956a) The Future of American Politics. Garden City: Doubleday & Co.

___(1956b) The Revolt of the Moderates. New York: Harper & Brothers.

MAY, GEORGE. (1955) "Iowa politics." The Palimpsest 32 (September): 250-252.

MURPHY, DONALD R. (1960) "The elusive farm vote." The Progressive 24 (October): 27-29.

NEPRASH, JERRY ALVIN. (1932) The Brookhart Campaigns in Iowa, 1920-1926. New York: Columbia University Press.

SMITH, T. LYNN. (1942) "The role of the village in American rural society." Rural Sociology 7 (March): 10-21.

U.S. NEWS AND WORLD REPORT. (1955) "How farmers really voted in '54." U.S. News and World Report 38 (January 7): 37-39.

WALLACES' FARMER AND IOWA HOMESTEAD. (1958a) "Issues in the election." Wallaces' Farmer and Iowa Homestead (September 20): 52-53.

___(1958b) "Elections won by 'shifters.' " Wallaces' Farmer and Iowa Homestead (December 6): 24-25.

___(1959) "How many farm Democrats?" Wallaces' Farmer and Iowa Homestead (January 17): 26-27.

___(1960) "Voting on November 8." Wallaces' Farmer and Iowa Homestead (November 19): 17.

Chapter **IV**

THE ACCOMMODATION OF INTERESTS

The relatively decentralized nature of the political system and the absence of effective party competition perhaps have enhanced the importance of interest groups in Iowa politics. While the dominant party has been reluctant to promote causes that lack proven popularity, the Republicans also have sought to accommodate political interests and goals that may contribute to their electoral majorities. The process has provided the major party with substantial benefits, but it has enabled private associations to partially replace political parties as the groups responsible for the organization and leadership of political movements.

The weakness of the Democrats, on the other hand, apparently has encouraged competing interests to form their own groups rather than to present their proposals through the minority party. Interest groups have scored major gains by attracting sufficient support to commend their ideas to the dominant party, but there has been little advantage in carrying programs to the opposition. As a result, most major conflicts in Iowa politics such as urban-rural disagreements have been conducted through organized interests rather than through the parties.

Interest groups usually organize and present political demands to government. In the absence of imposed party restraints, they are capable of performing many of the normal functions of political parties by developing and leading popular support for issues in the electorate and in governmental institutions such as the legislature. The delegation of party duties to interest groups, however, can yield both advantages and disadvantages for the political parties.

In Iowa, the Republicans have been deprived of a measure of party control, but they have enjoyed the advantages of embracing numerous politically mature organized interests as well as other benefits. Since political conflict frequently has erupted among opposing interests rather than between competing parties, the Republicans have been able to protect themselves in part from the internal strains often created by interest group demands. The willingness of the dominant party to relegate battles regarding significant issues to the affected interests not only has permitted the Republicans to avoid the repercussions of choosing between conflicting interest group objectives, but it has perhaps also inhibited the development of divisive factionalism within the state and the party. Although party direction may have been weakened in the process, the surrender of some partisan functions to organized interests in Iowa apparently has strengthened the dominant party at the polls.

The significant role performed by interest groups in Iowa politics perhaps has hindered the development of party competition along urban-rural lines. Since the Democrats usually have constituted a weak partisan alternative and the Republicans have sought to encompass both urban and rural demands, historically neither political party has been conspicuously identified with the interests of the two areas. In particular, the ability of Republicans to avoid the endorsement of positions that have been the objects of intense conflict has reduced party intervention in contests between urban and rural interest groups.

In the absence of firm partisan direction, competition between urban and rural groups frequently has produced some unique struggles and alliances. Since the parties generally have been unwilling or unable to exploit interest group disagreements among opposition segments of the electorate, conflict between organized interests seldom has reflected a clear urban-rural separation. On many occasions, coalitions have been formed between apparently divergent urban and rural organizations. Yet, in Iowa, the division between urban and rural groups has been a more prominent feature of interest group activity than any other type of distinction.

Since Iowa traditionally has been a rural state, farm organizations usually have occupied a position of major importance in political developments. Perhaps the peculiar intensity of agricultural politics also has increased the importance of interest groups in Iowa politics. At various times, militant organizations of farmers have arisen to improve their social and economic positions. Iowa always has been one of the first states to join such movements. Although agricultural groups often have eschewed political methods at their inception, they usually have soon resorted to politics to gain their objectives.

In part, the explanation for both the unusual propensity of farmers to form cohesive organizations and their reliance on political tactics may be related to the unique role of agriculture in the economy. Unlike other segments of the society such as labor and management, for example, farmers are highly vulnerable to the fluctuations in a market over which they can exert little influence or control. In addition, the increasing support provided agriculture by government programs has tended to make farm income largely a matter of political decision. As a result, farmers probably have more incentives to form organized interests than other segments of the population (Key, 1958:35).

The first farm organization to achieve political prominence in Iowa was the Patrons of Husbandry. The "first real farmer Grange was founded at Newton, Iowa, on April 17, 1868." By 1874, Iowa led the nation in the number of local Granges (Ross, 1951:97). In 1873, when the group was reorganized under the direction of farmers rather than the government officials who founded it, Dudley W. Adams of Waukon was elected the first national Master of the Grange.

Although the Grange was formed originally as a social or fraternal organization, the opportunities available through political action rapidly became apparent. In many chapters in Iowa (Buck, 1920:30), for example, "the restriction in the constitution of the order as to political or partisan activity was evaded by the simple expedient of holding meetings 'outside the gate,' at which platforms were adopted, candidates nominated, and plans made for county, district, and state conventions." In 1872, Adams (quoted in Kramer, 1956:39) charted the course that the Grange eventually was to follow when he proclaimed, "What we want in agriculture is a new Declaration of Independence. . . . We have heard enough, ten times enough, about the 'hardened hand of honest toil." The explicit statements in his speech hardly exceeded the prohibition on political activity, but the implications of his remarks were obvious to members of the organization and to other state politicians.

The political activity of the Grange was directed largely at correcting some of the abuses that farmers suffered at the hands of the railroads. The tactics of the organization included intense campaign efforts to elect members and other candidates sympathetic to the Grange position to public office as well as vigorous lobbying in the state legislature. Although Governor Cyrus Carpenter and a majority of Iowa legislators in 1872 were prominent members of the group, the Grange never was able to achieve most of its political aims. In 1874 the Grange was successful in securing the passage of the so-called "Granger Law" to regulate railroad rates, only to see the act repealed a few years later when the strength of the organization had ebbed.

Although the Grange suffered a dramatic loss of membership throughout the nation after 1874, an organization bearing its name but lacking its initial political strength continued to survive in Iowa. Perhaps the effort that the Grange originally focused on the railroad problem undermined its effectiveness on subsequent political questions. When railroad reforms were temporarily adopted, the political energies of the Grange seem to have been largely spent. Since there appeared to be no permanent agricultural issues to sustain interest in the organization, the Grange did not remain an important force in Iowa politics.

In any event, the sudden loss of membership after 1874 rendered the Grange politically impotent. In modern times, the organization seems to have become little more than a relatively unimportant political subsidiary of the larger and more powerful Iowa Farm Bureau Federation. In 1960, for example, the Master of the Iowa State Grange (quoted in Wiggins, 1963) commented, "We admit that the Iowa Farm Bureau is much larger than the Grange and you might say we support many of their actions."

The numerous efforts that were made by rural or farm organizations to curb the political power of railroad interests in the late nineteenth century were joined by an unusual continuity. Significantly, the leadership for many of the early agrarian movements in Iowa was furnished by agricultural editors who sought to improve farm prosperity rather than by farmers themselves. Among the journalists were such men as "Father" Coker F. Clarkson of the *Iowa State Register* and William Duane Wilson of the *Iowa Homestead*, both of whom were active in organizing the Iowa Grange; "Uncle Henry" Wallace of *Wallaces' Farmer*, both a grandfather and a father of United States Secretaries of Agriculture; B. F. Gue of the *Iowa Homestead*, a railroad commissioner under Governor Larrabee; Seaman A. Knapp of *The Western Stock Journal and Farmer*, an early leader in agricultural extension work; and "Tama Jim" Wilson, U.S. Secretary of Agriculture for seventeen years.

One of the first farm groups that Clarkson promoted was the State Agricultural Society which sponsored state and county fairs. However, this organization soon appeared to men like Wallace as "a fake farmers' 'front' for the railroads" (Lord, 1947:90). At a state convention of the Agricultural Society, a group of dissidents including Wallace, Clarkson, Wilson, and Knapp sat down at a well-known oyster supper to form the Agricultural Editors' Association.

In 1881 the organization of agricultural editors was instrumental in organizing the Farmers' Alliance in Iowa. Clarkson was the first secretary of the Alliance; and Newton B. Ashby, Wallace's son-in-law and editor of the *Farmer and Breeder*, became a national lecturer (Ross, 1951:108). Although Iowa belonged to the supposedly non-political "Northern" wing of the Alliance, it gave strong support to several candidates, including Governor Larrabee, who sought to reduce the influence of the railroads.

Perhaps the most important political activity of the band of agricultural editors, however, was the founding of the Farmers' Protective Association which fought the barbed-wire monopoly in Iowa. The Protective Association received major backing from other farm groups including the Grange and the Alliance as well as financial assistance from the state legislature. Although the farmers received a number of unfavorable court decisions, the organization finally succeeded in forcing the trust to lower the price of barbed-wire (Lord, 1947:95). As counsel for the Farmers' Protective Association, Albert B. Cummins launched a long and significant career in Iowa politics that included the curtailment of the influence of the railroads.

The struggle between rural or farm organizations and the railroad interests in Iowa constituted perhaps the principal source of urban-rural conflict in the late nineteenth century. Through the distribution of free passes and other favors, the railroads established a uniquely influential organization of professional men and prominent community leaders that was centered primarily in urban settlements. Despite the determined opposition of a succession of agrarian movements, the railroad interests aided by lobbyists such as General Dodge, J. W. Blythe, and Judge Hubbard compiled an enviable record in the Iowa legislature. In 1886, for example, "forty-five anti-railroad bills were defeated in the House and sixteen in the Senate, while the few that did pass were all modified and weakened in later years" (Nye, 1965:66). As the progressive faction within the Republican party gained increasing support, however, the railroads were forced to modify their tactics after the deaths of Blythe and Hubbard. In 1912 railroad spokesmen conceded that they had "made a mistake in the past by requiring their attorneys to appear for them" in the legislature, and they formed a committee of railroad men to serve as lobbyists (Des Moines *Register and Leader*, December 17, 1912). The powerful organization of professional men that had been created in nearly every city in Iowa was disbanded. As a result, new sources of disagreement replaced the battle between farmers and railroad or business interests as a focus of urban-rural disputes in Iowa politics.

Perhaps "a high point" in the growing consciousness of urban-rural differences was reached in the gubernatorial campaign of 1916 (Schmidhauser, 1963:2). In that year, urban leaders who were interested in educational reorganization and the systematic development of hard surface highways formed an organization called the Greater Iowa Association. Their proposals were vigorously promoted by an official of the group, E. T. Meredith of Des Moines, who was also the Democratic candidate for governor. Equally firm in his opposition to the measures was the Republican nominee, William Lloyd Harding, who espoused the rural position on the issues. Residents of rural areas supported small rural schools sentimentally and because of their convenience in sparsely populated districts. In addition, they championed increased expenditures for the thousands of miles of farm roads in the Iowa countryside to get "the farmer out of the mud." Since the latter program diverted funds from city streets and major thoroughfares, it provided a particularly intense and persistent source of urban-rural conflict.

The strength of rural voters was demonstrated not only in the election of 1916, which resulted in the largest Republican gubernatorial victory up to that time, but also in the passage of a referendum proposal for a constitutional convention in 1920. The convention plan was promoted by two relatively new farm organizations, the Iowa Farmers Union and the Iowa Farm Bureau Federation, that sought an amendment to prevent farm strikes or withholding actions from being declared unconstitutional. While the state legislature subsequently failed to establish the procedures for holding a convention, the political strength of organized rural groups clearly was revealed in the referendum victory, which represented the only occasion at which Iowa voters have approved a decennial call for a constitutional convention. Ironically, forty years later, one of the organizations, the Farm Bureau, was instrumental in defeating a proposed constitutional change that might have yielded increased representation for urban areas in the legislature.

Although the Farmers Union and the Farm Bureau originally shared the objective of promoting agricultural interests, their subsequent courses of action in Iowa politics have diverged considerably. Perhaps the more radical of the two groups was the Farmers Union, which received a state charter in Iowa in 1917 with an initial membership of 5,000. A leading figure in the organization was Milo Reno, who served as state president during the period of a rapid decline in agricultural prosperity from 1921 to 1930.

The relatively poor economic conditions of the twenties and the subsequent depression spawned a number of militant and aggressive farm movements. On March 19, 1931, for example, 1500 farmers stormed the legislature in Des Moines demanding repeal of a state law for compulsory tuberculin tests of cattle. Despite the efforts of the Farmers Union and other groups to mediate the conflict, Governor Dan Turner was compelled to summon National Guardsmen and impose martial law to quell the so-called "Cow War" that erupted near Tipton a few months later. Only under the force of arms did some Iowa farmers submit to the compulsory testing that threatened, however vaguely, to deplete their herds in the darkest hours of the depression.

At the state convention of the Farmers Union in September, 1931, the fiery Reno turned from usual programs for agricultural relief to advocate the formation of a new group and a new tactic for increasing farm income. In the summer of 1932, Reno became chairman of the Farm Holiday Association which proposed a total strike by farmers and the withholding of agricultural products from the markets until prices improved. Supported by a dedicated band of farmers, many of whom had followed him in the Farmers Union, Reno called for a general farm strike on October 21, 1933. Although the efforts of the Farm Holiday Association to enforce the market boycott provoked considerable controversy and some violence, the organization was not noticeably successful in achieving its purposes. Farmers had not yet attained sufficient unity to sustain the popularity or the effectiveness of the holiday program. As economic conditions improved, the movement gradually subsided.

Although Reno provided uniquely dynamic and vigorous leadership at a critical point in the development of Iowa agriculture, the strength of the Farmers Union was substantially undermined by radical causes. By organizing a new group to promote the farm strike, Reno probably drew more members and activity from the Farmers Union than he had attracted to it previously. The Farmers Union in Iowa also has been troubled by charges of Communist infiltration. During the so-called "red scare" of the twenties and again after World War II, such accusations frequently were leveled against the organization. While the attacks did not destroy the group, the prestige and effectiveness of the Farmers Union suffered from the charges.

The relative prosperity of Iowa farmers and the loss of leadership to other movements has provided the Farmers Union in Iowa with a more limited political role than it has enjoyed in many neighboring states. Unlike the Grange, however, the Iowa Farmers Union has consistently acted in opposition to the larger and more influential Farm Bureau organization. Although the membership of the Farmers Union has been confined to rural areas, the position of the group has been increasingly favorable to the Democratic party and to urban or labor objectives. In 1960, for example, the Farmers Union endorsed the proposal for a constitutional convention to grant increased legislative representation to urban areas. Despite its unique position, however, the Farmers Union has not achieved the status of a prominent group in Iowa politics.

In 1955, another in the series of farm organizations appeared in Iowa. As a group that sought to alleviate farm problems by withholding products in order to negotiate contracts for improved prices, the National Farmers Organization seemed to be a spiritual descendent of the referendum campaign for a constitutional convention in 1920 and of the Farm Holiday Association. Ironically, one of the founders of the NFO was former Governor Dan Turner who had called out troops to squelch the so-called "Cow War" in 1931.

Like most prior farm movements, the NFO originally sought to avoid active participation in politics. By concentrating on the enrollment of a sufficient number of farmers to ensure the success of withholding actions and on bargaining for contracts to increase farm prices, the organization hoped to secure improvements in agricultural income without resorting to appeals for governmental or political assistance. Although the NFO experienced the same spectacular initial growth in membership that has marked other farm movements, it evidenced some tendencies to become involved in political controversies, possibly in association with Democratic or labor groups. In 1963, for example, the NFO was instrumental in supporting a so-called "red meat check-off" bill to deduct a small amount of money from market sales for the promotion of Iowa beef products. Although more legislators reported that they had been visited by constituents supporting the bill than by any other group except the representatives of rural elective cooperatives and the Farm Bureau, the measure remained buried in committees throughout the legislative session. Despite growing indications of political activity, therefore, the NFO has not been an influential group in urban-rural controversies in Iowa.

The most powerful and significant farm organization in Iowa, the Iowa Farm Bureau Federation, arose in response to still another agrarian movement. In 1918, Republican

Governor William L. Harding, who had been elected with strong rural backing, became worried about plans of the Non-Partisan League to extend its sphere of influence from the Dakotas to Iowa. Since the League had fomented a successful revolution in North Dakota by creating an independent political organization, Republican politicians naturally were anxious about the potential effects of the new movement on their rural sources of support. As a result, Harding contacted leaders of the newly formed Farm Bureau Federation and encouraged them to conduct a whirlwind organizational campaign starting in the northern tier of counties and working south, just ahead of League organizers (Kile, 1948:62). The success of the Farm Bureau membership drive meant that the Non-Partisan League in Iowa "never got a good head start" (Morlan, 1955:203).

The program for agricultural relief promoted by the Farm Bureau seemed to provide a basis for agreement among widely divergent segments of the population. From December, 1918, to December, 1919, the Farm Bureau increased from 39,600 to 104,000 members. By 1920, there were 135,000 Farm Bureau members in Iowa (Kile, 1921:125). While the Bureau offered no extreme solutions to the farm problem, it supported a number of governmental measures that enjoyed extensive popularity. Perhaps the theme of the new group was sounded in the keynote address of J. R. Howard of Iowa, first national president of the Farm Bureau, at the organizational conference of the American Farm Bureau Federation in 1919, when he said (quoted in Kile, 1921:118): "I stand as a rock against radicalism, but I believe in an organization which strikes out from the shoulder." Significantly, E. T. Meredith, who had espoused an urban platform in the gubernatorial election of 1916, also was among the delegates to the first organizational meeting of the national Farm Bureau.

Unlike many other farm groups, the Farm Bureau immediately assumed an active role in Iowa politics. In 1920, for example, Farm Bureau membership lists were employed to defeat senatorial candidate Smith W. Brookhart in the Republican primary as well as to secure the passage of the proposal for a constitutional convention. Furthermore, "the Bureau's numerical strength enabled it in 1920 to force both political parties in Iowa to include in their platform an agricultural plank drawn by the Bureau" (Schou, 1960:86). Perhaps the most significant political accomplishment of the Iowa Farm Bureau, however, was its role in organizing the so-called "farm bloc," a bipartisan coalition of Midwestern legislators in the U.S. Congress devoted to the support of measures to improve farm income. Plans for the "farm bloc" were first developed at a meeting of Iowa Congressmen and Farm Bureau officials in Des Moines on November 13, 1920 (Shideler, 1957:69). While most of its early ventures into Iowa politics were successful, the Farm Bureau subsequently has confined its role in major state campaigns to elections which seemed to threaten its effectiveness as a group.

In 1948, the Bureau loaned its membership lists and organization to state Senator William Beardsley in a successful effort to defeat incumbent Governor Robert Blue in the Republican primary. Blue had aroused the enmity of the organization by opposing a Bureau program for state financing to improve rural roads without increasing property taxes. Since 1948 represented one of the few primaries in which an incumbent governor has been rebuffed in his bid for party renomination, the active endorsement of the Farm Bureau apparently can have a significant impact on voting in Iowa. After the election, Beardsley said privately that the support of the Farm Bureau assured any state candidate of at least ten percent of the farm vote in Iowa (*The Iowan*, 1956:11). While this figure probably reflected a very conservative estimate, few politicians have been prone to ignore a guarantee of even one in ten votes.

After Governor Beardsley took office in 1949, the legislature established a road use tax fund to distribute 50 percent of the money to rural farm-to-market and secondary roads, 42 percent to primary roads, and 8 percent to municipal roads. Despite the urban-rural inequities in the system, the state legislature supported by the Bureau perpetuated the formula for a number of years. In 1961, for example, urban leaders organized an intensive but unsuccessful movement to secure a "fair fifteen" percent of the road funds for city streets and highways. Clearly, the allocation of road funds in particular as well as other forms of state assistance proposed by the Bureau and other rural groups has been a major impetus for urban-rural friction in Iowa.

The rural influence of the Farm Bureau also was enhanced for many years by the quasi-governmental function that it performed in supporting agricultural extension services which sought to acquaint rural families with modern and economical farming practices. From 1918 to 1955, Iowa law directed counties to pay up to $5,000 to agricultural associations with at least 200 members who contributed $1,000 annually in dues to assist extension work (Hardin, 1952:38). The separation of the extension service and the Farm Bureau was forced by a U.S. Department of Agriculture policy in 1955, but the Iowa Farm Bureau sought to ensure that it would not be replaced by a rival organization aiding in extension services (Block, 1960:226). Although the two organizations were formally divorced, traces of sympathetic cooperation between the Farm Bureau and the extension service continued to be evident. In 1961, for example, the Bureau adopted the only exception to its general policy of firm opposition to property tax increases when it endorsed legislation to augment the maximum amount that counties could collect in property taxes to finance extension work (Wiggins, 1963:30). After 1955 the Farm Bureau could no longer maintain its official alliance with the U.S. extension service in Iowa, but the long association between the two groups undoubtedly attracted members and prestige to the Bureau in rural areas.

Another issue on which the Farm Bureau has sought to exert its influence in state campaigns concerned the reapportionment of the state legislature. In 1960, the Bureau was instrumental in defeating a proposed constitutional convention that might have increased urban representation in the legislature; but it suffered a major setback in 1963 when a Bureau sponsored reapportionment plan was defeated in another state referendum. The energy expended by the Farm Bureau in preventing a reapportionment that would enhance the strength of urban areas perhaps reflected not only its interest in preserving rural domination but also its concentration on political influence in the legislature. As a focal point in Iowa politics, the legislature has maintained a uniquely significant role in the formulation of state policy. Similarly, as perhaps the most powerful interest group in Iowa, the Farm Bureau has devoted its political work primarily to shaping policy in the state legislature.

The most effective tool of the Farm Bureau in Iowa politics has been the sympathetic support of members of the legislature. While a "farm bloc" was organized in the U.S. Congress, a "Farm Bureau bloc" was born in the Iowa General Assembly. Since 1922, more than ten percent of the elected representatives in each session of the legislature, except for the meetings held during the New Deal years, have listed Farm Bureau membership in their official biographies. Significantly, this statistic reached its peak in 1963, at the height of urban-rural tensions, when nearly 40 percent of the legislators recorded Farm Bureau membership. While such figures clearly underestimate the actual number of Farm Bureau members in the legislature, a relatively high proportion of legislators considered their memberships in the Farm Bureau of sufficient social or political significance to list it in their biographies. With a large number of potential allies in the legislature, the explanation for the acknowledged effectiveness of the lobbying activities of the Farm Bureau has not been difficult to imagine.

In addition, many leadership positions in the Farm Bureau have been held by legislators, and many legislative leaders have been connected with the Farm Bureau. In 1955, for example, 28 percent of the legislators who listed Farm Bureau memberships also mentioned that they had been former county presidents of the organization. Similarly, a disproportionate number of committee chairmanships have been controlled by Farm Bureau legislators. In 1959 every committee in the Iowa House, except one, was chaired by a Farm Bureau member. Rural dominance in the leadership positions of the legislature probably has been strengthened not only by the system of apportionment but also by the organized activity of groups such as the Farm Bureau.

Further evidence of the influence of lobbying activities by the Farm Bureau was provided by a 1963 survey of all 158 members of the Iowa legislature. When they were asked which groups or organizations outside the legislature often have been "listened to" on major state issues, 58.9 percent of the senators and representatives cited the Farm Bureau. This was nearly 30 percent more legislators than referred to the next most frequently mentioned interest group. More than 90 percent of the legislators were able to identify Harry Storey, the principal lobbyist for the Farm Bureau. Furthermore, 32.9 percent acknowledged that they received "the most useful information" from Farm Bureau lobbyists. Fifty-seven percent of the legislators also remembered that they had been visited by Farm Bureau delegations from their constituencies during the legislative session, and 22.1 percent recalled that they had met with the Farm Bureau in their home counties. On nearly all questions that attempted to measure the association between interest groups and legislators, the Farm Bureau was mentioned more often than any other state organization.

The Farm Bureau has formulated a rather intricate procedure for bringing the weight of its membership to bear on political issues. Each year legislative meetings are held by local groups, and "opinionnaires" are distributed to all Farm Bureau members in Iowa. After the "opinionnaires" have been collected, the official policy of the state organization is drafted by the House of Delegates, a body of active Bureau members. Despite the efforts to involve a large number of members in the policy-making process, in 1959 only about 16 percent of all Iowa Farm Bureau members returned the "opinionnaires" or attended local meetings on legislation (Wiggins, 1963:7). As in most organizations, therefore, the policy of the Farm Bureau probably has been determined by a relatively small group of active participants.

The Iowa Farm Bureau Federation also has developed perhaps the most "completely organized" system for encouraging close contact between legislators and influential local advisors to enforce the policies of the organization (Truman, 1958:326). In each legislative district, a five-man Farm Bureau Legislation Committee has been formed to act as a liaison between the legislator and the local organization. Members of the committee must be strong supporters of Farm Bureau policies, locally prominent politicians, members of the same political party as the legislator chosen in the last election, and preferably active workers in his campaign. In addition to their advantages for gaining the confidence of legislators, the requirements for committee membership have been designed to provide a prominent position for potential candidates to unseat the elected representative, if he should bolt the Farm Bureau position on major issues. As a Farm Bureau organizing official candidly admitted, "We make sure that at least one member of the committee can run against the legislator and beat him, and we make sure that the legislator knows this." During each session of the Iowa legislature, members of the county Farm Bureau Legislation Committees also make at least one trip to the state capitol to visit their elected representatives. The annual visits clearly have yielded the respect and co-operation of many legislators, including one rural Republican who commented, "The Farm Bureau comes down every year. Other groups come only as the bills come out. That's what makes the Farm Bureau so much more powerful." While the Farm Bureau has seldom endorsed or supported candidates financially, the organization probably has been able to exert more influence on legislators by creating committees that reflect the personal and partisan associations of successful candidates than by supporting nominees who are subject to the possibility of defeat at the polls.

Although the influence wielded by the Farm Bureau might have been expected to provoke serious opposition from business or other urban groups, substantial conflict between business and agricultural organizations has not been a major feature of urban-rural tensions at least in the twentieth century. In fact, the generally conservative character of Farm Bureau policies often has attracted the enthusiastic support of business interests and produced a working partnership that some critics have termed an "unholy alliance" between agricultural groups and major business organizations. Since the Farm Bureau long has been the dominant rural interest group in Iowa, urban-rural controversies have been materially altered by the relationship between the Bureau and business interests.

Perhaps the leading business lobby in Iowa politics has been the Iowa Manufacturers Association. Although this organization has not acquired a large membership, the prominence of its members as well as its financial resources and prestige have made it a potent force in the state legislature. When legislators were asked in 1963 which interest groups were most frequently "listened to," 28.5 percent mentioned the Iowa Manufacturers Association; and 12.7 percent admitted receiving "the most useful information" from its lobbyists. In addition, 79.7 percent of the legislators were able to identify Harry Linn, Executive Secretary of the Iowa Manufacturers Association and a former state Secretary of Agriculture. As an important interest group, the Iowa Manufacturers Association was ranked second after the Farm Bureau by the members of the state legislature.

On most major issues involving urban and rural considerations, the accusation frequently has been made that the Iowa Manufacturers Association and the Farm Bureau have formed an alliance to work with the Republican party. In referenda campaigns of 1960 and 1963, for example, the IMA and the Farm Bureau united in their opposition to a reapportionment of the legislature that might have increased the representation of urban areas. Significantly, the Iowa Manufacturers Association was one of the first groups to endorse the Bureau-sponsored apportionment scheme that would have ensured the perpetuation of a rural majority in the legislature. As one rural state senator commented, "The Farm Bureau and the Iowa Manufacturers Association at their state conventions came out for this thing, which leads me to believe that there was a coalition."

Charges regarding an alliance between the Iowa Manufacturers Association and the Farm Bureau have not been confined to the issue of legislative apportionment but they have also been alleged concerning taxes, business regulation, and general state policy. In 1963, for example, an urban legislator claimed that the Iowa General Assembly was controlled by "the marriage of the Iowa Manufacturers Association with the Farm Bureau and all those who come under their sway—and that's the majority." Despite the predominantly urban base of manufacturing interests, the alleged coalition between the Farm Bureau and the IMA often has been regarded as a rural combination that opposed urban aspirations. In 1961 Democratic State Chairman Lex Hawkins (quoted in Des Moines *Register*, May 8, 1961) even charged that the Iowa Manufacturers Association and the Farm Bureau were impeding the process of urbanization by working together to prevent "big industry" from entering the state because it also would bring " 'big labor,' [which] would mean a swing to the Democratic party." While the coalition probably has not been as powerful as many critics have claimed, the joint activities of the principal agricultural and business groups in Iowa undoubtedly have tended to reduce urban-rural conflict.

Allegations regarding the alliance of the Iowa Manufacturers Association and the Farm Bureau perhaps have developed from the fact that the two organizations share numerous political interests and objectives. The relative lack of a broad popular following, however, probably has made the Manufacturers Association the weaker partner in the so-called coalition. While the IMA frequently has contributed to the attainment of the political goals of the Farm Bureau, the Bureau alone often has seemed to possess sufficient influence to secure official sanction for many of its policies. As one perceptive member of the capitol press corps observed, "For all practical purposes, the Iowa Manufacturers Association now is the tail on the Farm Bureau dog."

Perhaps the third most important lobbying group in Iowa has been another business organization, the Iowa Taxpayers Association. In the 1963 survey of all Iowa legislators, 13.9 percent reported that they received "the most useful information" from the Taxpayers Association, and 59.5 percent were able to identify Ray Edwards, the principal lobbyist for the group. The complexity and political importance of tax issues has give the Taxpayers Association, as a group that has concentrated on the fiscal activities of state government, a unique position of influence in the legislature.

Although the two organizations have been technically distinct, the Iowa Taxpayers Association and the Iowa Manufacturers Association usually have reflected not only common policy objectives but also similar memberships. As one prominent Republican legislative leader commented, "I think if you compare the membership lists of the Iowa Taxpayers Association and the Iowa Manufacturers Association, you will find that they are pretty much similar." Since the business interests represented by the Manufacturers Association also have tended to predominate in the Taxpayers Association, the two organizations have shared policy positions both with each other and with the Farm Bureau. The three major interest groups in Iowa politics often have formed a coalition of agricultural and business interests that has hindered the development of urban-rural conflict.

Perhaps the only organization that has assumed a consistent and distinctly urban position on state issues is the Iowa Federation of Labor, AFL-CIO, which also has been among the weakest interest groups in Iowa. In 1963 only 3.8 percent of the legislators said that they had received "the most useful information" from labor organizations, and only 32.3 percent were able to identify Charles L. ("Vern") Davis, state president of the AFL-CIO and a principal labor lobbyist. Unlike both business lobbies and other groups, the legislative access of labor has been limited to a relatively small band of members, consisting primarily of urban representatives and Democrats.

While the relative impotence of labor in Iowa politics often has been ascribed to its association with the minority party, labor also has not been totally divorced from the Republicans. Ray Mills, long-time president of the Iowa Federation of Labor, did not switch his party affiliation from Republican to Democrat until 1954. In some cities in Iowa, unions have continued to endorse and support local Republican candidates. The militancy and effectiveness of labor probably has been modified by its relationships with the political parties. While organized labor never has been able to deliver a decisive vote to the Democrats, many Republicans have resented what they regarded as the unrealistic endorsement of state Democratic candidates by labor. As a result, labor has been forced to play a rather passive role in most major political controversies in Iowa.

One issue on which labor has attempted to exercise a major influence is the so-called "right-to-work" law to prevent the negotiation of agreements that would make union membership compulsory. Despite the presence of thousands of union members in the state capitol, the legislature passed a "right-to-work" bill in 1947. Labor leaders subsequently contended that the law prevented them from enrolling sufficient members to become an important political force in Iowa during a critical period of industrial expansion. Others, however, claimed that the intransigent opposition of labor to the measure limited their political impact in a predominantly rural legislature. As one Republican party leader noted, "At every session of the legislature, our majority leaders sat down with the labor boys and asked them what they wanted. They always said, 'repeal of the right-to-work law,' and we always pointed out that that was impossible, and that was as far as the discussion went." For a considerable time, therefore, labor experienced numerous frustrations due to the weakness imposed by the inability to achieve its major political objective.

As urbanization has increased, however, labor gradually has assumed a more aggressive position in promoting urban interests. In 1963, the state president and secretary-treasurer of the Iowa Federation of Labor were the plaintiffs in a suit that successfully challenged the constitutionality of the apportionment of the legislature. In addition, labor played a prominent role in the defeat of the Farm Bureau-sponsored reapportionment plan in a state referendum. Although organized labor has encountered numerous obstacles in its struggle for political distinction in Iowa, the victories seemed to lend some support to the hope that the increased representation of urban areas might strengthen the role of labor in the state legislature.

Traditionally, the influence of major interest groups in Iowa politics has been related to the attainment of their objectives. As one rural Republican legislator observed, "There are king-makers in any state; the question is how well they can make kings. When I first came to this state, the Iowa Manufacturers Association ran the legislature. Then the teacher's association, in Charlie Martin's hay-day, had a powerful influence. Now the Farm Bureau seems to be on top." In large measure, the strength of important mass membership organizations has seemed to correspond with their needs. During the period after World War II, for example, business interests sought and received certain laws, including the "right-to-work" measure, to curtail the growing power of labor. Subsequently, the Iowa State Education Association pressed for the reorganization of school districts and finally attained its objectives, despite rural opposition. Finally, the Farm Bureau attempted to secure property tax relief and to maintain the predominance of rural members in the legislature. Although most of the major battles by influential interests in the Iowa legislature have contained overtones of urban-rural conflict, perhaps the most significant controversy has concerned reapportionment, since it could determine the basic urban and rural composition of the principal arena in which nearly all policy questions are resolved.

In addition to the dominant interests that have shaped important state policies, a large number of specialized organizations have had a decisive influence on the outcomes of relatively minor squabbles that have not contained profound urban-rural implications. While such groups have lacked sufficient membership to participate in electoral campaigns, they often have been represented in the legislature by experienced lobbyists who possessed a virtual monopoly on information regarding bills of particular interest to them. In addition to their expertise, the lobbyists for relatively small organizations frequently have relied upon legislative friendship and fellowship to reach their goals. A prominent example has been the Crowley brothers. For many years, L. E. ("Roy") Crowley has served as Executive Secretary of the Iowa Motor Truck Association, while his brother, Orville, has represented the Associated General Contractors of Iowa. Speaking of "Roy" Crowley, one legislator marveled, "When he puts on a party, he really puts on a party. He flipped this [truck] length bill over in the Senate overnight. He's got the moxie to do it." Although some lobbyists for the relatively small groups have worked almost exclusively with the committees which have jurisdiction over their interests, others have been known and respected by a large number of legislators.

Since experienced professional lobbyists possess important and vitally needed information about specialized legislation, they normally encounter little difficulty in obtaining access to the legislators. One Republican representative assessed the various interest group representatives in the following terms:

> There are three types of lobbyists. You have these men who are executive secretaries of associations. You can get

> material from them pro or con. You can depend on them because they're here year after year. You have lobbyists who are paid to kill a particular bill. You have to watch their information, because they're just interested in the money. Then there are the do-gooders. They are neither effective nor dependable.

The expertise and continuity of the representatives of relatively small interest groups probably have made them among the most influential types of lobbyists in the Iowa legislature.

As the practice of lobbying has become more complex, however, a new breed of lawyers who represent groups in the legislature much as they would clients in court has arisen. In 1963, for example, a Des Moines attorney, Edward H. Jones, lobbied for eight organizations including such major interests as the Iowa State Bar Association, the Iowa Bankers Association, and the Iowa State Educational Association. Jones also was cited by 17.7 percent of the legislators as the lobbyist who provided them with "the most useful information." In the same session, another capitol city lawyer, George A. Wilson, son of a former Republican Senator from Iowa, appeared in behalf of seven interest groups. The need for legal assistance in drafting legislation as well as the expense of maintaining a permanent office apparently has inspired many groups to retain an attorney rather than a full-time employee to represent them before the legislature.

The techniques that lobbyists have employed to secure the passage or defeat of legislation have been as numerous and as varied as the circumstances with which they must deal. A common device, however, has been the provision of dinners and entertainment for legislators. Although some members of the legislature have questioned the propriety of interest group hospitality, there has been little evidence that such privileges have affected the fate of legislation unduly.

The widespread bribery and corruption that was typical of many state legislatures in an earlier age apparently has been replaced by the distribution of small favors that probably foster friendship and understanding more than they sway votes.

On occasion, lobbyists also have resorted to a ruse or deception in order to further their legislative aims. Since the division between urban and rural interests often has been blurred, groups sometimes have tried to capitalize on the confusion between rival organizations. In 1872, as an example, four railroad lobbyists attempted to pose as farmers visiting the legislature. The failure of their disguises was revealed by a Senate resolution that stated in part:

> WHEREAS, there have been constantly in attendance on the Senate and House of this General Assembly . . .four gentlemen professing to represent the great agricultural interests of the State of Iowa, known as the Grange; and–
>
> WHEREAS, these gentlemen appear entirely destitute of any visible means of support; therefore be it–
>
> RESOLVED, By the Senate, the House concurring, that the janitors permit aforesaid gentlemen to gather up all the waste paper, old newspapers, etc., . . . and that they be . . . given a pass over the Des Moines Valley Railroad with the earnest hope that they will never return to Des Moines [Buck, 1920:53-54].

In more recent years, a favorite device of some lobbyists has been to sit at the back of the House chamber at the beginning of each session and to watch the votes as they are recorded on the electric tally board. Particular attention is paid to freshmen who hesitate momentarily in pushing the button on the assumption that the new legislators are following the cue of another member. In this way, the lobbyists seek to gain some knowledge of the networks of influence in the legislature and of the representatives who might be singled out for special treatment and attention. Since legislative voting agreement in Iowa commonly has reflected a personal following rather than the influence of political parties, the technique apparently has been an effective one for lobbyists.

The failure of a lobbying ruse in 1963, however, illustrated the folly of completely ignoring political parties in the legislative process. After a bill to establish a Fair Employment Practices Commission had been log-jammed in legislative committees by a major professional organization, an attorney retained by the group responded to the special pleas of its sponsors by drafting a similar bill that would make employment discrimination a criminal offense. According to a prominent Republican party leader, the counsel probably drew up the bill so that it would be particularly objectionable to enough members of the Senate Sifting Committee to keep it permanently bottled up. On the assumption that the version was acceptable to the economic interest group, however, it was adopted by a Republican caucus and passed in the House. When the bill reached the Senate, the majority leader reported that he was having trouble getting the legislation out of the Sifting Committee and asked a Republican party official for help. The official deliberated with the committee until one of its members finally said, "Well, if this is a party measure, I guess we'd better pass it." As a result, the bill was quickly approved by the Senate and signed by the Governor. When the Republican leader returned to his office, he was greeted by an angry call from the group counsel who complained bitterly about the meddling of the party that had caused his plan to backfire.

The likelihood that the Republican party will intervene in the legislative process, however, normally has been as remote as the possibility that the legislature will be totally deceived by the tricks of a lobbyist. As the party leader who intervened in behalf of the equal employment opportunities bill acknowledged, "Until I came into office, I doubt if a Republican party official had ever set foot inside the legislative chambers." Similarly, in 1963, few legislators reported that they were visited by or received information from party organizations. In general, the parties in Iowa seldom have attempted to influence legislation, and legislators have not expected interference from the parties.

The withdrawal of the dominant party from legislative battles apparently has facilitated the solution of major controversies in Iowa politics. Since lobbies have expressed the relatively limited and specialized interests of their members, they have been able to compromise legislative goals more easily than the parties which must somehow embody the aspirations of the entire electorate. As a result, political disagreements in Iowa frequently have been resolved by affected interest groups, and the absence of party interference has removed a major obstacle to the fulfillment of organizational demands.

The important role performed by the lobbies in reaching necessary legislative compromises was exemplified by another case of an issue that involved major state interests and far-reaching significance at a time when urban-rural tensions perhaps were at a peak. For many years, public and private utilities in Iowa were allowed to operate with a minimum of state supervision. In 1963, however, a move was made to place the utilities under the regulation of the Iowa Commerce Commission. As a result, 32.9 percent of the legislators reported that they had received "the most useful information" from utilities lobbyists. After months of discussion and negotiation, a compromise finally was achieved between the two major contestants in the controversy, the utilities companies and the Iowa League of Municipalities. When the bill reached the House floor shortly thereafter, the chairman of the Public Utilities Committee rose so frequently to indicate which amendments were part of the compromise and which were not that the Speaker was moved on occasion to remark humorously: "The gentleman will be happy to learn that he has permission to adopt his amendment." With the quiet acquiescence of the House, the bill was quickly approved by the Senate and the Governor, and Iowa obtained the first regulation of utilities by a commission in its history. Although the total ratification of interest group agreements has been relatively rare, there has been little doubt that the compromises of lobbyists have had a significant impact on state policy.

The tendency to legislate through compromises reached by affected interest groups has been a common feature of Iowa politics. Perhaps the absence of intense party competition has encouraged the parties to withdraw from controversies that might alienate segments of the electorate. The consequent decentralization or abdication of normal party functions has resulted in the diminution of party discipline or control. In part, this development has stemmed from a natural reluctance to take a position in conflicts between groups of voters that have been traditional supporters of the dominant party. The Republicans have been unwilling to risk the estrangement of friendly groups by exerting firm direction in the reconciliation of opposing legislative goals. The party has also been able thereby to reduce the possibility that competition between interest groups might disrupt or upset the prevailing Republican tradition in Iowa. Differences have been resolved more satisfactorily by the affected interests than between or within the parties. As a result, the one-party legacy has seldom been threatened by divisive cleavages between organized interests. At the same time, the costs to centralized party authority have been great.

The minority party, on the other hand, generally has been willing to accept groups that might contribute to its efforts to build a majority coalition in Iowa. Since Democrat support traditionally has offered few prospects for success, however, major organizations usually have attempted to avoid identification with the opposition party. Consequently, partisan divisions seldom have had a major effect on the compromises reached by conflicting interest groups.

Perhaps more significant than the role of lobbies in compromising opposing interests, however, has been the cooperation between various organizations. In large measure, the success or failure of major legislative proposals has been shaped by the agreements formed between interest groups. As a veteran rural Republican legislator commented, "We have some effective lobbying groups, the Bar Association, the Bankers Association, the Farm Bureau, the teachers lobby in times past, the Iowa Manufacturers Association, and the Motor Truck Association. Those are the ones who often pool their resources for a single objective. They trade influence. It is a ricocheted type of influence that is most effective." While many of the major interests have wielded considerable influence individually, their combined power has been enhanced significantly.

Although the major organizations in Iowa politics generally have reflected urban-rural distinctions, the relationships between the groups frequently has been characterized by cooperation and collaboration rather than by conflict. During the late nineteenth century, antagonism between farm and railroad interest was a marked feature of Iowa politics; but, since that time, the political aims of agricultural and business organizations often have been united. In fact, perhaps the only organization that has opposed the predominantly rural objectives of the principal interest groups in the modern era has been the relatively weak and ineffectual Iowa Federation of Labor. As a result of the close working arrangements between the principal agricultural and business lobbies, the tensions that might have developed from the distinct geographical organization of major groups traditionally have failed to flourish in Iowa.

By forming coalitions or compromises on significant state policies, however, the interest groups have played a critical role in Iowa politics. The resolution of controversies by the affected organizations perhaps has increased the likelihood that major interest group demands will be satisfied and correspondingly reduce the incentives for the participation of lobbies in factional or partisan contests. Since the dominant party has been reluctant to intervene in divisive controversies, the prominent role of interest groups in Iowa also has enhanced the importance of the legislature in the political system.

REFERENCES

BLOCK, WILLIAM J. (1960) The Separation of the Farm Bureau and the Extension Service. Urbana: University of Illinois Press.

BUCK, SOLON J. (1920) The Agrarian Crusade. New Haven: Yale University Press.

HARDIN, CHARLES M. (1952) The Politics of Agriculture. Glencoe: The Free Press.

THE IOWAN. (1956) "The politicians look at the election." The Iowan 5 (November): 11-13.

KEY, V. O., JR. (1958) Politics, Parties, and Pressure Groups. New York: Thomas Y. Crowell Co.

KILE, ORVILLE M. (1921) The Farm Bureau Movement. New York: The Macmillan Co.

___(1948) The Farm Through Three Decades. Baltimore: The Waverly Press.

KRAMER, DALE. (1956) The Wild Jackasses. New York: Hastings House.

LORD, RUSSELL. (1947) The Wallaces of Iowa. Boston: Houghton Mifflin Co.

MORLAN, ROBERT L. (1955) Political Prairie Fire. Minneapolis: University of Minnesota Press.

NYE, RUSSEL B. (1965) Midwestern Progressive Politics. New York: Harper & Row.

ROSS, EARLE D. (1951) Iowa Agriculture. Iowa City: The State Historical Society of Iowa.

SCHMIDHAUSER, JOHN R. (1963) Iowa's Campaign for a Constitutional Convention in 1960. New York: McGraw-Hill Book Co.

SCHOU, JOHN T. (1960) "The decline of the Democratic party in Iowa, 1916-1929." Unpublished M. A. thesis, State University of Iowa, Iowa City.

SHIDELER, JAMES R. (1957) Farm Crisis, 1919-1923. Berkeley: University of California Press.

TRUMAN, DAVID B. (1958) The Governmental Process. New York: Alfred A. Knopf.

WIGGINS, CHARLES W. (1963) "The politics of the Iowa Farm Bureau Federation." Unpublished paper, Washington University, St. Louis, Missouri.

Chapter 

THE INSTITUTIONAL CONTEXT

Although informal associations such as parties, factions, and interest groups generally form the bases for major conflicts within a state, political divisions also are reflected in the institutions of government. Since formal sources of authority represent the focus or objective of political activity, prominent groups usually seek either to impose their designs on the governmental apparatus or to adjust their strategies to patterns which the structure of government imposes upon them. In the process of adaptation, the legal machinery of government gradually becomes modified by significant cleavages in the state.

In addition, political strife often has tended to institutionalize important sources of conflict. In Iowa, the stress between farmers and business interests and between urban and rural areas has had a profound impact on state government. While the governor and other state officials have been compelled to rely upon all areas of the state in the execution of similarly expansive responsibilities, members of the state legislature generally have served as the spokesmen only for the individual districts that elected them. In a state where controversies frequently follow geographic lines, the natural tendency of the legislature to mirror critical contests for political power has provided it with a significant role in state government.

During most of its history, the Iowa legislature has been the haven of rural or small town Republicanism. Except for the New Deal years, Republicans controlled a majority in the Iowa General Assembly throughout the twentieth century until 1964. Frequently Republican dominance was so overwhelming that the Democrats were unable to offer effective or significant opposition. In 1953, for example, the three Democratic members of the House of Representatives could literally have "caucused in a phone booth." Furthermore, the legislature seldom has exercised the constitutional duty of altering its own institutional structure to reflect the growth of urban areas in the state. In part, the failure or refusal of the legislature to reapportion according to changes in the population may have been related to the desire of Republicans to maintain their predominant position. Generally, however, legislators of both political parties have been reluctant to tamper with the system of representation that placed them in office. Since each succeeding session of the legislature has reflected and inherited the rural characteristics that prevailed for many years after the state was founded, there have been few incentives to change the basis of representation.

Officials chosen by the state as a whole have been subjected to a self-adjusting mechanism that virtually precluded them from completely ignoring the voting strength of urban or rural areas, but most legislators from the malapportioned districts have been free to perpetuate rural dominance and to disregard the distribution of the urban and rural population in the entire state. The disproportionate influence of rural or small town interests undoubtedly has tended to subdue urban-rural conflict by impeding the consideration of measures that would be of principal advantage to urban residents. Furthermore, since state senators and representatives in Iowa have been elected by districts that do not cross county boundaries, the legislature has not reflected the tensions between cities, towns, and farming areas that often existed within counties. Although the Iowa legislature has assumed a prominent position in state government in part because it has embodied the separate interests of important groups more completely than officials elected by all voters in the state, the identification of basic cleavages has been complicated by the tendency of the legislature to personify rural or small town values and to obscure differences between urban and rural voters within counties.

Yet, the persistent struggles between the political interests of urban and rural residents have been reflected by changes in the composition of the Iowa General Assembly. In large measure, trends in the occupational characteristics of the legislature have denoted differences in the representation of urban and rural areas in Iowa. Although farmers and lawyers have not been chosen solely as the legislative representatives of their respective trades, they have tended to characterize distinct geographical bases and styles of life. Most attorneys have practiced law only within small towns or cities, and normally farmers have pursued their work exclusively in rural areas where they reside. Consequently, farmers generally have been associated with the representation of rural voters, while lawyers usually have been viewed as the spokesmen for urban or small town residents. An empirical investigation of legislators from urban and rural areas has tended to confirm this observation. From 1909 to 1963, for example, 71.4 percent of all senators and representatives from Polk county, the most populous area in the state, were lawyers. Some farmers have represented predominately urban counties in the legislature, but generally the election of lawyers and farmers has implied the representation of urban and rural areas, respectively.

Lawyers and farmers have always formed two of the principal groups in the Iowa legislature. In the period from 1899 to 1927, for example, it was found (Schaffter, 1929:40-42) that "in both Senate and House about forty-five percent of the members are either lawyers or farmers." Furthermore, the proportions of farmers and lawyers in the legislature have tended to fluctuate in relationship to each other. In the legislative sessions of 1899 and 1901, the proportion of lawyers in both the Senate and the House was "much larger than the average" for the era from 1899 to 1927, "and the proportion of farmers much smaller"; but, at the end of the period, the number of lawyers was "much below average and the number of farmers much above." The trends in the ratios of farmers and lawyers in the legislature indicated the periodic displacement of urban-based attorneys by legislators who lived on farms.

Figure 2 reveals the proportion of lawyers and farmers who served in each session of the Iowa General Assembly from 1909 to 1963. A large proportion of legislators in Iowa have been either lawyers or farmers. At every session during this period of more than fifty years lawyers and farmers together constituted a majority of all members of the assembly. In addition, the occupational attributes of the membership have tended to indicate the predominately rural character of the Iowa legislature. Only during World War II did the number of lawyers exceed the number of farmers in the House and the Senate.

Perhaps the most striking features of the graph, however, are the amazingly clear and specific patterns displayed by the percentages of lawyers and farmers. As the propor-

oportion of Farmers in Each Session of the Iowa Legislature, 1909 - 1963

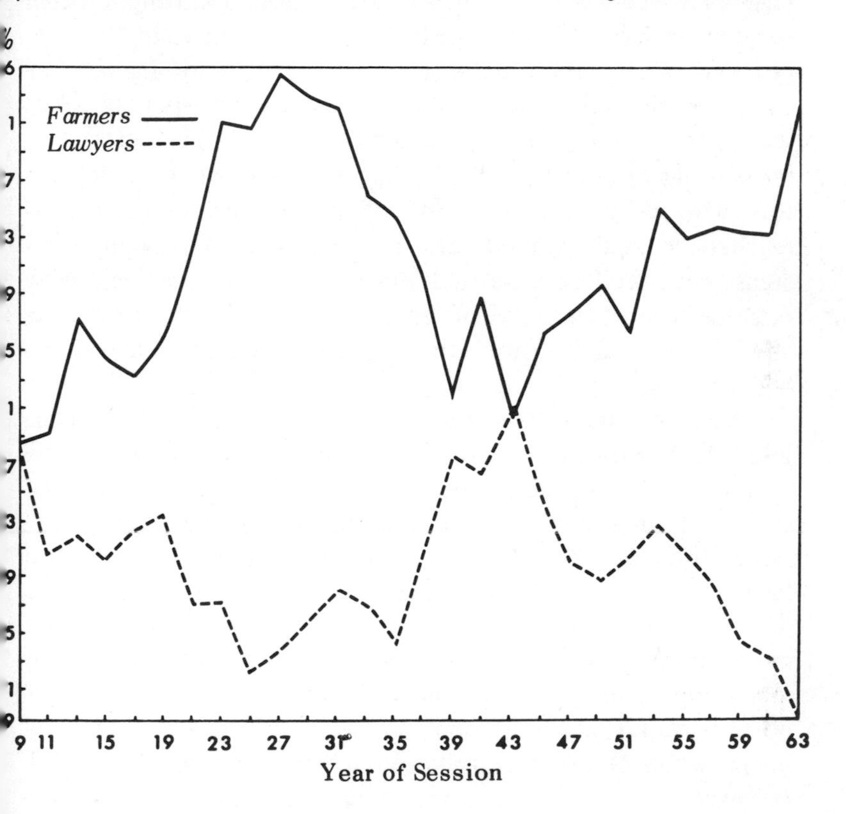

Figure 2

tion of farmers increased, the proportion of lawyers correspondingly declined, and vice-versa. The phenomena have reflected a decision by voters within counties to substitute farmers for lawyers as their representatives during critical periods of Iowa history. In large measure, shifts in the proportions of lawyers and farmers in the legislature have seemed to be related to the political activity of urban or rural segments of the county electorates. The largest group of farmers was present in 1927 when the proportion of legislators who were not residents of urban communities also reached a peak (Schaffter, 1929:46-48). Although rural areas were well represented during most of the legislative sessions, the percentage of farmers increased most dramatically at the same time that agrarian unrest reached its highest levels.

Perhaps the most feasible interpretation (Schaffter, 1929:43-44) of the rotation of farmers and lawyers can be found "in the character of economic opinion at the time of each particular choice of representatives by the electorate." Generally, the trends in the membership of lawyers and farmers in the legislature have seemed to mirror the cycle of agricultural prosperity. The proportion of farmers rose most abruptly in the late 1920s and 1950s, both periods of increasing economic stress in rural areas. Although the proportion of farmers declined somewhat during the New Deal years, when Democrats replaced Republicans as the majority party in the Iowa legislature, farmers continued to predominate as at least symbolic spokesmen of the demands for agricultural relief. Farmers usually have become most active in politics when they are faced with immediate economic crises. Apparently declining economic conditions also have tended to promote the selection of farmer representatives. As one investigator (Gold, 1961:85-86) has speculated:

> It may be that the lawyer's representation of diverse interest groups tends to be displaced by special interest group representatives when the special interest group faces a crisis and is seeking some form of legislative aid. In "good times" politics is left to the lawyer. Thus, to some extent, the heavy participation of lawyers may represent a lack of political motivation on the part of other occupational groupings.

In Iowa, however, lawyers who reside in urban areas seldom have been the major group in the legislature. Although there seem to have been more incentives to choose farmers as representatives during periods of economic adversity than during periods of prosperity in rural areas, farmers in Iowa have never lacked the motivation to prevent lawyers from dominating the legislature.

The preponderant influence of farmers undoubtedly has reduced the effective representation of urban residents in the Iowa legislature, but it also has tended to dilute the cohesion of rural members. Since urban spokesmen seldom have challenged the power of farmers or rural representatives, there has been little incentive for rural legislators to form a distinct and identifiable bloc in the legislature. As a result, few dissimilarities have been suggested by a cursory examination of voting by members of the legislature from urban and rural counties. An investigation (Price, 1959:80) of legislative voting in the three sessions from 1951 to 1955, for example, found that "in the Iowa General Assembly on 15 percent of all definite opposition roll call votes, a rural majority was opposed to an urban majority. It should be noted that approximately three-quarters of the votes taken in the General Assembly were votes without definite opposition. Therefore, of the total number of roll call votes taken in the General Assembly during the period, only 4 percent had a rural majority opposed to an urban majority. On only 26 votes, or less than 2 percent of 759 definite opposition roll calls, did two-thirds of the rural delegation oppose two-thirds of the urban bloc." The largest proportion of roll call votes in the Iowa legislature apparently reflected unanimity rather than conflict; and, even on measures that provoked some disagreement, there was little evidence of urban-rural antagonism.

A number of factors, however, indicate that the analysis of roll call voting by legislators from urban and rural counties may have obscured as much as it revealed about controversies in the Iowa General Assembly. Many potential sources of urban-rural opposition were cloaked by unrecorded votes on amendments, compromises reached by committees or interest groups, and even appointments to crucial committees. The study (Price, 1959:83) of legislative voting from 1951 to 1955, for example, admitted that since "a heavy majority of the legislature believed that one-quarter to one-third of the controversial votes remained in committee, the percentage [of urban-rural conflict] probably would have been somewhat higher if all potential urban-rural division votes had [been] reported out of committee." Since rural members usually have controlled an overwhelming majority in the Iowa legislature, contests between urban and rural interests seldom have been decided by public votes on the floor of the General Assembly.

The ability of rural representatives to determine committee assignments and other processes for the consideration of legislation has reduced the opportunities for urban-rural conflict and the necessity of a unified rural vote in the legislature. During the period from 1951 to 1955, the urban delegation voted with at least a two-thirds cohesion on 71 percent of the definite opposition roll call votes; but there were substantial defections by rural representatives from the consensus of their group. Significantly, rural Democrats were somewhat more likely to support the urban delegation on issues that produced rural-urban disagreements than urban Republicans. Although partisan affiliations and other considerations may have affected voting patterns, the relatively low cohesion of rural members may have been stimulated by the power of rural legislators to influence the outcome of legislation prior to a vote.

Furthermore, since the representation of most legislative districts in Iowa has been based on counties rather than population, few members have been elected by distinctly urban or rural areas. As the analysis of the occupational characteristics of Iowa legislators indicated, urban-rural cleavages may be more evident within counties than between counties. Even if there were a close relationship between population and representation, urban-rural differences might not produce distinctions in legislative voting. Some legislators from districts consisting primarily of large cities may have shared the beliefs and voting records of rural legislators, and vice-versa.

Conflict between urban and rural legislators also may have failed to produce significant differences in the voting behavior of the two groups. Although few controversies in the Iowa legislature have yielded intense urban-rural opposition, there have been substantial disparities in the support that urban and rural legislators have given many issues. In the legislative sessions from 1951 to 1955 (Price, 1959:54), only questions concerning reapportionment and the butter-oleomargarine controversy provoked major disagreements between rural and urban legislators, but "the urban bloc was likely to give 20 to 25 percent higher support than the rural delegation for increased social welfare and appropriations, more liberal labor and liquor laws, more money for education, . . . and a greater degree of local autonomy for the cities." Since the largest number of recorded votes in the legislature has been taken on technical or non-controversial matters, a relatively low proportion of urban-rural conflict in legislative votes may not have indicated the absence of disagreements between legislators from urban and rural counties. In order to determine the prevalence of urban-rural differences on roll call votes, both general patterns of legislative voting and votes on major issues that have produced extensive controversy must be examined.

In large measure, the voting behavior of members from urban and rural counties apparently has depended upon the type of issue that was presented to the legislature. In addition, conflict on some questions has intensified as particular items of legislation have become increasingly salient to urban and rural legislators. In the state senate from 1957 to 1961, for example, roll call votes revealed a greater increase in the cohesion of senators from districts containing a city of at least 10,000 population on business regulation than on any other issue (Berg, 1962). The nature of questions that produced growing distinctions in legislative voting behavior may have reflected the increasing importance of certain issues not only to urban and rural legislators but also to significant interests within the state.

One issue that always has been of profound importance to every major segment of the population of Iowa, including urban and rural residents, is the question of taxation. Since the enactment of a tax policy for the state has a significant impact on policital and economic resources, tax measures clearly provide a more accurate measure of differences in legislative voting than questions that produced little public interest and few important consequences. In 1963, the legislature sought to revise the tax structure of Iowa by adopting an omnibus tax bill. Although the proposal finally was defeated, a survey of all the members of the legislature revealed that 76.6 percent mentioned taxation as one of "the most important issues in this session of the legislature."

To determine the extent of urban-rural conflict on

significant legislative problems such as taxation, a series of 65 recorded and unrecorded roll call votes on the omnibus tax bill was analyzed from the 1963 session of the Iowa legislature. Generally, urban and rural members were classified on the basis of whether or not their districts contained a city of 10,000 or more population. Although the urban delegation included a slightly higher proportion of lawyers and Democrats than the rural group, 63 percent of the urban members were Republicans, 24 percent were farmers, and 43 percent were engaged in some occupation other than law. The major distinction between urban and rural legislators was related to the population of the districts they represented and not to additional factors that might have influenced their voting.

The results of the analysis (Hahn, 1965) disclosed that on slightly more than one-half of the roll call votes concerning the tax bill, in both the House and the Senate, an urban majority was opposed by a rural majority. Although the urban-rural schism was most pronounced on votes affecting the passage of the bill, it also was apparent on amendments concerning the sales tax, agricultural land tax credit or property tax replacement, and luxury taxes on items such as beer, cigarettes, trading stamps, and pari-mutual betting. Differences in the voting behavior of urban and rural legislators were evident on several other amendments, but roll call votes on major questions of taxation produced a clear division between opposing urban and rural groups in the legislature.

The investigation of roll call votes has indicated that there was a relatively small amount of urban-rural conflict on many questions, but differences between urban and rural members have been a prominent feature of voting on significant issues in the Iowa legislature. The distinction has been crucial because some proposals obviously were more important than others. Measures that have a significant impact on public policy have tended to reflect prominent cleavages in the state. Legislation of limited or minor importance, on the other hand, has failed to mirror clear differences in voting behavior. Since controversies have focussed on questions of wide interest and importance, the prevalence of urban-rural disagreement in the legislature on matters of critical importance has extensive implications for the outcomes of political contests in Iowa.

Even in 1965, after the legislature was reapportioned on the basis of population and Democratic members replaced Republicans as a result of the landslide in the 1964 election, urban-rural differences on roll call votes continued to play a conspicuous role in the Iowa legislature. Although there was a substantial increase in party voting during the session, urban and rural members were sharply divided in the amount of support given legislation endorsed by a majority of the members of their party. Democrats in both the House and the Senate with the highest party support scores on 416 roll call votes were legislators from urban constituencies. Conversely, rural Democrats recorded the lowest party support scores. There were virtually no urban Republicans in the House, but Republican senators from urban districts gave significantly less support to their party than rural Republican senators (Wiggins, 1965). Apparently the differences in party cohesion revealed by the voting records of urban and rural Democrats in the House and by rural and urban Republicans and Democrats in the Senate reflected major divisions between urban and rural objectives. As the parties increasingly began to serve as the spokesmen for distinct rural and urban groups, however, significant new trends were developing in the legislature. Since the Democrats were sustained principally by the votes of urban legislators and the Republicans viewed their principal support as coming from rural members, the voting patterns of rural Democrats and urban Republicans displayed substantial evidence of party irregularity. The tendency of the parties to represent separate urban and rural interests, therefore, diminished rather than increased the impact of party discipline on significant issues in the legislature. As long as basic cleavages between urban and rural residents in major areas of public policy remained a viable force in Iowa politics, neither political party could completely overcome the effects of urban-rural differences on legislative voting.

Although the analysis of voting and occupational characteristics provides a broad picture of the legislative branch of Iowa government, it fails to yield much information concerning the internal structure of the legislature. Rural representatives and farmers usually have enjoyed a numerical advantage in the Iowa General Assembly, but general characteristics must be related to the daily operations of the Senate and the House. In particular, further analysis is required to determine the effect of urban-rural cleavages on significant sources of leadership in the legislature.

In the legislative session of 1963, a period of rural Republican dominance, all 158 members of the legislature were asked to name the representatives or senators who were most influential in the passage or defeat of legislation (see Hahn, 1970). In the House, only one of the twelve representatives mentioned most frequently as legislative leaders was from an urban area. Members recognized as powerful legislators from rural counties included two former Speakers, a former minority party leader, the majority leader, the current Speaker and Speaker pro tem., and the chairmen of four important committees. The most acclaim, however, was received by Dewey Goode, a veteran Republican representative from a rural county, who was cited by 40 percent of the members of the House, or twice the proportion that mentioned any other legislator. Although Goode did not hold a major position of formal leadership in the Iowa General Assembly, he was widely respected as an authority on Iowa statutes and as a principal advocate of rural interests. Several rural legislators commented that "Dewey Goode knows the Code of Iowa better than most lawyers." In a large and diffuse political body, experience and expertise have been primary qualifications for leadership.

Although most of the influential representatives in the House were rural farmers or merchants, four of the most frequently mentioned leaders in the Senate were lawyers from relatively large cities. According to the survey, the leading Senator in the 1963 session was David Shaff, a young Republican attorney from Clinton who had become somewhat identified with rural interests because of his sponsorship of the Shaff plan for legislative apportionment. But three of the remaining eleven leaders cited by members of the senate were lawyers who have been prominent spokesmen for urban residents.

Perhaps the contrasts between leadership in the House and the Senate reflected traditional differences in the two bodies. Historically, there have been fewer farmers and rural representatives in the Senate than in the House. During the period from 1899 to 1927, for example, the Senate contained about twice as many lawyers and urban residents as the House (Schaffter, 1929:40-41). Since senators often have been elected by multi-county districts based partially on population instead of single counties, urban legislators have gained a somewhat greater voice in the Senate than in the House. Although the constitutional requirement that bills must pass both houses to become law and other factors have prevented the Senate from completely overcoming rural dominance in the Iowa legislature, urban senators have had more opportunities to rise to positions of prominence and authority than urban members of the House.

In both the Senate and the House, most of the informal leaders mentioned by their colleagues had served in major legislative offices. In 1963, all the members of the Iowa legislature were asked to rank the importance of formal officers of the legislature including the Speaker or the Lieutenant Governor, the Speaker pro tem. or the President pro tem., the Majority Leader, the Minority Leader, the Chief Clerk or Secretary of the Senate, the chairman of the Steering Committee, and the chairman of the Sifting Committee. The Steering Committee and the Sifting Committee have been appointed in nearly every session to screen the bills introduced in the legislature. The Steering Committee in the House has determined the order in which bills receive attention on the calendar, and the Sifting Committees in both chambers have selected the measures to be considered during the closing days of the session. All of the officials, except the Minority Leader, have been chosen by the majority party in each house.

There was little doubt among the legislators about the influence of the principal officers. The Speaker or the Lieutenant Governor was ranked first by 58.2 percent of the members, and 63.9 percent rated the Majority Leader in one of the top two positions. Similarly, the relatively insignificant role of the Minority Leader was recognized by a large proportion of the members of both parties. The Minority Leader was placed in one of the bottom three positions by 70.8 percent of the legislators.

The influence of the Speaker and the Lieutenant Governor undoubtedly has been a product of their legal authority to appoint the committees, assign bills, and preside in the House and Senate, respectively. Although the Speaker is selected by a caucus of the majority party and the Lieutenant Governor is chosen in a state election, both offices generally have rotated among men who have had prior service in the legislature. Since the presiding officers usually have not retained their positions long enough to accumulate personal power, the leadership exerted by the Speaker or the Lieutenant Governor has stemmed more from the power of their office than from individual authority.

If it were necessary, however, to select the most influential single figure in Iowa politics, the Lieutenant Governor might be a logical candidate for that position. Since he has not been elected by members of the legislature, the Lieutenant Governor has more discretion and fewer obligations in the appointment of committees. As a Republican state senator commented:

> If you would get a Lieutenant Governor that would stay in several years and be hard-fisted, he would be the strong man in the state. He is the most powerful man in state government. He can stop any piece of legislation here, and he can do it in a way so that no one knows he does it. He can put the squeeze on all the state departments. He appoints the chairmen of appropriations, and they report to him periodically.

In 1959, when a Democrat was elected to preside over a Republican controlled Senate, the power to appoint senate committees was removed from the Lieutenant Governor. The authority apparently was considered sufficiently important to require partisan control. In analyzing the structure of Iowa government, the crucial role of the Lieutenant Governor also was acknowledged by a former Republican legislator and United States Senator who said, "One of the most powerful offices, if not the most powerful, is the Lieutenant Governor. You won't get any legislation through unless he says it's O.K."

Since the preeminent roles of the Speaker and the Lieutenant Governor have been based in large measure upon their ability to appoint committees, clearly the members and chairmen of key committees also exercise disproportionate power in the legislature. The fate of most bills introduced in the legislature usually has been decided in committees. As a result, debate and votes in the chamber often are perfunctory, and most members acquiesce in the recommendations of the committees. As one urban Republican representative commented, "We have a kind of Gresham's Law that operates in the legislature. The good stuff is hoarded. The things that circulate are grandiose emotions." Because much of the relevant information necessary for the construction of effective arguments frequently has been in the exclusive possession of committee members and lobbyists, legislators have been forced to rely upon the judgments of members of committees in whom they have confidence. Members of committees with jurisdictions over particularly significant legislation, therefore, have acquired a substantial amount of influence in the Iowa legislature.

In 1963, the members of the Iowa General Assembly also were asked to select the committee that "most of the members would like to be on if the choice were up to them." The top choice of the legislators was the Appropriations Committee, which was cited by 41.1 percent of the House and the Senate. Only 7.6 percent chose the Sifting Committee, and 17.7 percent picked Appropriations and Ways and Means. The preferences of the legislators were confirmed by several former Speakers who reported that for many years between 80 and 95 of the 108 members of the House had requested appointments to the Appropriations Committee.

The popularity of the Appropriations Committee apparently was related both to the desire to bequeath money to favored institutions and to the fact that the recommendations of the Appropriations Committee seldom have been challenged in the legislative chambers. As one Republican legislator advised, "If it's a session when you have money to give, get on a subcommittee that represents a group you favor." The entire budget for all agencies of state government normally is prepared by appropriations subcommittees in both the House and the Senate. In the final days of a legislative session, the reports of the Appropriations Committees customarily receive the speedy and nearly unanimous approval of the legislature. Since the pressures of time impose severe limitations upon budgetary controversies, the decisions of the Appropriations Committee generally have received more deference in the legislature than the conclusions of other committees.

The power of the members of critical committees, the Speaker, the Lieutenant Governor, and other legislative leaders has depended in large measure upon the willingness or ability of the officials to exercise the authority that they have been granted by legislative rules or statutes. As one of few leaders who are informed about the work of the committee, the Speaker or the Lieutenant Governor, for example, can have a significant effect upon the deliberations of the Appropriations Committee. Similarly, in the assignment of bills, the presiding officers often can determine the fate of legislation. Much, however, depends upon the extent to which they choose to exercise those powers. As a rural Republican senator remarked concerning the Lieutenant Governor, "If he's easy, he'll say to the Secretary of the Senate, 'You refer the bills.' If he's strong, he won't do this." The delegation of authority to minor officials may enhance the popularity of the elected officials in the legislature, but it also can contribute to the dispersion of control in the Iowa Assembly.

Variations in personal willingness or ability to exercise influence may account for differences in the importance of legislative offices. In the survey of all members of the legislature, for example, some representatives and senators, in ranking the influence of major officials on legislation, indicated slight confusion about the roles of several legislative leaders. Three officials, the Chairman of the Steering Committee, the Chairman of the Sifting Committee, and the Chief Clerk or Secretary of the Senate, received ratings in all seven categories. The Speaker pro tem. or the President pro tem. was ranked in sixth or seventh place by 48.3 percent of the legislators, but another 20.4 percent rated them in third position. The question revealed, however, that the Chief Clerk or Secretary of the Senate received the most votes for third-place, the Chairman of the Sifting Committee got the most fourth-place votes, and the Chairman of the Steering Committee obtained the most fifth-place votes. As ad hoc committees charged with the responsibility of deciding which legislation will be submitted to a vote on the floor, the Steering and Sifting Committees exercise a determinative influence over a broader range of bills than any other group in the Iowa legislature. Yet, as many legislators pointed out, the influence of the committees generally has depended upon the supervision and direction that they have been given by the Speaker or the Lieutenant Governor.

Several factors have tended to promote the influence of committee chairmen, presiding officers, and other leaders on the operations of the legislature. Initially, the Iowa General Assembly only has met biennially for 100 day sessions that customarily adjourn at the beginning of the corn planting season. As a result, legislators usually have lacked the time required for a careful study of each bill, and they have been compelled to rely to a considerable extent on the statements of other legislators and lobbyists who possess sufficient information to guide their opinions.

In addition, the traditionally high turnover of members has tended to deprive the legislature of needed information and experience. During the period from 1925 to 1935, for example, 49.4 percent of the legislators in Iowa served one term, and only 21.9 percent remained in office for three or more terms (Hyneman, 1938). Apparently, most of the legislators who served for short periods were retired voluntarily rather than forcibly from the legislature. Between 1925 and 1933, only 20.3 percent of the members of the House and Senate were defeated in primary elections. In part, the brief tenure of senators and representatives may be related to the lack of adequate compensation for legislative service. Iowa legislators traditionally have been paid $30 per day for each day that the legislature is in session, or an average of approximately $3,000 during a two year term. Most legislators have not been able to sustain their expenses or the losses from their usual occupations on the salaries, and legislative business often has been left to the few members who can afford the costs of returning repeatedly to the legislature. Since there has been little incentive for continuous service, influence naturally has concentrated in the small group of senators and representatives who have acquired a familiarity and an understanding of legislative operations through seniority.

The legislative process often has been employed by the leaders to further their own policy objectives. In particular, the appointment of committees and the assignment of bills have been favorite devices for promoting or quashing legislation. A rural Republican representative recalled, for example, that J. Kendall ("Buster") Lynes, former majority leader of the Senate, spent "the first couple weeks in the session in finding sure 'graveyards' for certain pieces of legislation he did not like. This amounted to picking a subcommittee which would sit on the bills all session long."

Although the appointment of committees has provided legislative leaders with the opportunity to exercise a dominant influence over legislation, there has been little controversy in Iowa concerning committee assignments. Unlike many one-party states, the minority party often has been denied committee chairmanships in the Iowa legislature. A solution to the problem was suggested by one Democratic state senator in 1963 who commented:

> There are twelve Democrats in the Senate, and I have always felt that we should have one-fourth of the committee chairmanships. But they don't give the Democrats any recognition. I spoke up against it. Then one day I got four committees. It's the squeaky wheel that gets the grease.

Few similar complaints, however, have been raised concerning the denial of important committee positions to urban legislators. A study (Hyneman and Carey, 1954:26) of representatives who were elected by the most urban and the most rural areas in the state from 1925 to 1953 revealed that members "from the most urban districts held eleven chairmanships for every twenty seats they occupied in the House; legislators from the rural districts held eight chairmanships for every twenty seats."

Although a careful examination of committee assignments and chairmanships has suggested that urban members have occupied fewer prominent or critical committee positions in both the Senate and the House than rural representatives, criticism of the lack of urban representation on important committees has been limited for a number of reasons. Urban members generally have acquired less seniority than rural legislators. They have been more likely to be Democrats than their rural colleagues. Perhaps most significantly, however, the failure of urban legislators to obtain crucial committee positions has been a natural product of the numerical superiority of rural members in the House and the Senate. Since urban areas usually have been underrepresented in the legislature, they also have been underrepresented on significant committees. There have been no efforts in Iowa or elsewhere to apportion committee members or chairmen on the basis of population. Similarly, farmers generally have been more successful in acquiring committee chairmanships than lawyers. In the House, farmers held slightly less than half of all chairmanships from 1925 to 1953; in the Senate as well, farmers occupied more committee chairmanships than any other group (Hyneman and Carey, 1954:40). The traditionally rural character of the Iowa General Assembly has been reflected in committee positions.

The system of leadership in the Senate and House, however, has been closely related to urban-rural cleavages. Members of important committees, chairmen, and other leaders have considerable opportunity to wield a decisive influence on legislation, and committee leadership normally has been determined by the appointments of the presiding officers. Consequently, controversies between urban and rural legislators in the House have tended to revolve about the selection of the Speaker. A survey of all members of the legislature revealed that the basic conflict in the election of the Speaker of the House traditionally has involved, as the Speaker himself put it, "the rural Farm Bureau group versus the other groups." Discussions with several former Speakers and prominent party leaders outside the legislature confirmed the fact that the election of a speaker at the beginning of each legislative session usually has pitted farmers and rural representatives against members from urban areas and their allies. In addition, rural legislators generally have won the contest for the speakership. Only 3 of the 27 Republican Speakers from 1900 to 1963 were from urban counties.

Since the Speaker has considerable authority over the committees and the legislative process generally, the results of a session often have been determined by the choice of a presiding officer. While infrequent victories by the urban faction may have increased urban-rural tensions, the success of rural representatives in the party caucus has reduced measurable or overt conflict in the legislature by limiting the consideration of legislation that would arouse urban-rural antagonisms or that would principally benefit urban areas. On some occasions, the leadership and operations of the legislature have suggested the relative absence of significant controversies in the Iowa General Assembly. Yet, at the basic point of political power, urban and rural interests have marked the visible division between members of the majority party.

The choice of the presiding officers in the Senate by the entire state electorate rather than by a group of legislators has complicated and limited the identification of urban-rural differences in the upper chamber. Yet, more than three-fourths of the Republican Lieutenant Governors of Iowa have come from rural areas. Despite the somewhat urban tendencies of the Senate, therefore, rural areas usually have been represented by the presiding officers of both houses. As a result, the predominately rural composition of the legislature has been reflected in the leadership structure of both houses of the Iowa General Assembly.

The significance of the speaker and the lieutenant governor in Iowa government has derived not only from their influence in the legislative process but also from the fact that a large proportion of the governors of the state have been recruited from the two positions (Schlesinger, 1957:29-30). Between 1912 and 1950, for example, six of the eleven men who served as Governor of Iowa had been former Lieutenant Governors. "The importance of legislative leadership . . . is shown by the fact that many of those who became governor without becoming lieutenant governor did have some form of legislative leadership. In Iowa, two men were speakers of the house." Furthermore, most of the Lieutenant Governors who became Governor previously had held a major position in the legislature. "Only one of Iowa's lieutenant governors had held no previous office. The remaining five had been in one or the other house of the state legislature, and two had been speakers of the house." A common pattern of succession in Iowa has extended from the legislature to speaker to lieutenant governor and finally to the office of governor.

In large measure, however, "the office succession hierarchy in Iowa" has been "a post-1912 phenomenon" (Schlesinger, 1957:42). Prior to 1912, the selection of Republican governors generally was determined by the outcomes of factional struggles between railroad or business interests and their rural opponents. The suggestion has been offered (Schlesinger, 1957:42) that the line of succession has developed since 1912 because "the conflict within the party has been concentrated upon contests for national office, especially upon the senatorship, thus leaving the lesser offices by default to the Republican machine."

In part, conflicts over the selection of major federal representatives from Iowa have been related to the choice of governors. The adoption of the Seventeenth Amendment, which removed the election of U.S. Senators from the state legislature, however, did not entirely divorce the contests for the two offices. In 1912, for example, the intense factional squabbles in the legislature over the selection of a Senator to fill the vacancy caused by the death of Senator Dolliver apparently determined the succession of Iowa governors for nearly 20 years. A former Republican state chairman, whose father was a prominent member of the legislature in that session, explained that four legislative leaders "were assured of their succession to the governor's chair by the deals that were made in that election." Nearly all of the Iowa governors who served from 1912 to the depression in 1932 were the leaders of major factions in the session of the legislature that chose a successor to Senator Dolliver. The establishment of the custom of elevating former leaders of the legislature to the office of governor probably represented an effort to mitigate factional conflict within the legislature rather than a willingness to relinquish control over the governorship by concentrating on national offices.

The formal pattern of succession in Iowa has tended to minimize gubernatorial competition within the Republican party. While the tradition of succession was developed by the compromise in the Senatorial election of 1912, it was immediately re-established when the Republicans returned to power after the depression ebbed in 1938. Undoubtedly many aspiring candidates for governor have been discouraged because they did not possess sufficient experience to qualify for the office. As an urban legislator commented on the tendency to recruit legislators for higher office, "One of the main reasons is that we come from a one-party state. In a one-party state, it's a slow, tedious task to work your way up."

In addition, the crucial role of the legislature in the Republican party has given legislators an important advantage in the race for major state offices. In 1963, all 158 members of the Iowa legislature were asked for their interpretations of the fact that 20 of the last 24 governors and lieutenant governors of Iowa were former members of the legislature. The interest of the candidates in politics was mentioned by 23.4 percent of the members, and 20.9 percent cited their experience. An additional 16.5 percent claimed that service in the legislature gave candidates a better background to discuss state problems, and 15.8 percent said that a legislative record was a means of becoming known to the voters. Perhaps more significant, however, were the responses of 15.2 percent who called the legislature a natural stepping stone, and 13.9 percent who pointed out that candidates for higher office from the legislature have a potential nucleus of support in each county of the state.

When Iowa legislators were asked directly whether or

not "the legislature is looked upon as a stepping stone to other political offices," the replies indicated that membership in the General Assembly frequently has been a vehicle for political advancement. Although some legislators reported that service in the legislature often has marked the end rather than the beginning of a political career, 46.2 percent felt that the legislature generally was regarded as a stepping stone to higher office and 34.2 believed that some members looked upon it that way. In addition, 76.6 percent of the members said that they had helped candidates secure nominations for elective offices, and 65.2 percent reported that they had worked for other legislators in primary contests for major state positions. Most Iowa legislators have not refrained from promoting the candidacies of their colleagues in competition for nominations to higher public offices.

The support that legislative candidates for higher office receive from fellow legislators undoubtedly has been a significant factor in many Republican nominations. A rural Republican representative explained, for example, that most governors and lieutenant governors have been former legislators "because they have an opportunity to be known to fellow legislators who are . . . important party people in their respective districts. Thus, they have a nucleus of support. Legislators trust them. Being human, they dream." The centers of support in each county provided by legislative friends and colleagues has been a major asset in numerous state campaigns. Since most Iowa counties have been represented by Republicans, legislative candidates for higher office in the dominant party have had the opportunity of extending their networks of influence throughout the state with their contacts in the legislature. As another Republican legislator observed, "Acquaintances in the legislature give you a start toward a state organization. In every county you have some key personnel you can go to. You'd be surprised at how many people ask their senator or representative who to vote for."

The tendency of the leaders and members of the House and Senate to become major state candidates has indicated that the Iowa General Assembly not only has reflected major controversies between urban and rural interests but also that the legislature itself has operated as a significant faction within the Republican party. Republican legislators generally have maintained the cohesion and organizational loyalties that normally have distinguished major factions within a political party. Most members expected the legislature to produce several contenders for important offices, and they accepted the responsibility of actively promoting the campaigns of fellow legislators in primary elections. The results of their activity have revealed the strength of the legislative branch of the Republican party. In addition to the high proportion of Iowa governors who attained the position only after extensive leadership and service in the legislature, candidates for nominations to other offices who have been supported by the "legislative faction" had enjoyed a substantial advantage over aspirants recruited from outside sources. As a prominent Republican member of the Iowa congressional delegation acknowledged, "The reason I'm here more than any other thing is my legislators."

The crucial position of the legislature within the Republican party has had an important impact on urban-rural controversies. By encouraging the elevation of legislative leaders to the governorship, the pattern of succession has tended to project the results of urban-rural differences in the selection of leaders in the General Assembly to the highest levels of Iowa government. Furthermore, since the legislature traditionally has represented the rural areas of the state, the advancement of legislative candidates to the office of governor has yielded some disadvantages for urban aspirants. In the process of serving and leading the party within the legislature, potential governors have a substantial opportunity to absorb prevailing rural interests and objectives. While gubernatorial candidates must appeal to both urban and rural residents in state-wide elections, values and experiences acquired in the legislature cannot be totally discarded. During much of the twentieth century, therefore, the formal lines of succession to the governor's chair in Iowa have tended to reduce urban-rural conflict by discouraging the candidacies of urban politicians outside the legislature and by promoting the predominately rural interests of the "legislative faction" within the Republican party.

In addition to the custom of recruiting major state officials from leadership positions in the state legislature, the selection of Iowa governors has been influenced by other factors. "Although there is no public discussion of . . . geographical rotation in Iowa, if a line is drawn just below the 42nd parallel, the state will be divided into northern and southern areas which have alternated in providing governors with one exception until the Democrats began to win in the thirties" (Schlesinger, 1957:30). The tradition of rotating governors between northern and southern Iowa probably was inspired by differences in the economic resources of the two areas and by continual friction over the explosive issue of taxation. As a leading rural Republican representative from northern Iowa commented:

> I would separate Iowa into three categories: northern Iowa, southern Iowa, and the urban areas. Northern Iowa has the greatest agricultural potential. Southern Iowa doesn't have enough assessed valuation. We have to be a big brother for them. This is also true of the cities. We, in our area, would be better off with no state funds. The urban areas need the help, but they don't contribute to it proportionately. Polk County has the lowest number of assessments in the state.

Although the division of urban and rural tax funds has been a matter of considerable controversy, the less productive farm lands of southern Iowa probably have contributed less to state tax coffers than the more advantaged rural areas in the north. The rotation of Republican gubernatorial candidates probably balanced opposing interests that developed from contrasts in economic resources and undermined the bases of potentially disruptive conflict concerning taxation. While the clash of interests between northern and southern Iowa apparently was obscured by the emergence of the Democratic alternative, the battles between urban and rural representatives over tax revenues and allocations possibly may eclipse most prior forms of political conflict in Iowa.

The increasing success of Democratic candidates for governor in capturing both the attention and support of Iowa voters seemingly has had an impact upon the pattern of succession within the dominant party. Several years ago, one observer (Schlesinger, 1957:43) suggested that "the development of a Democratic party with real hope of success at the state level might have the effect of drawing out of the Republican party most of the disruptive forces and thus permitting the promotional hierarchy to be firmly established." However, increased party competition apparently has shifted rather than fixed the scheme of succession in the majority party.

Since the initial post-war Democratic gubernatorial victory in 1956, three of the first four Republican candidates for governor had been former attorneys general. As a rural Republican state senator noted, "There's been a change in the last few years. I took it for granted, and I think the people did too, that the governor should come from either house, a former Lieutenant Governor or a former Speaker. Then a couple of Attorneys General took a stand to enforce liquor and other things. I think it would be better for cooperation if the governor came from the legislature."

The tendency of Republicans to promote the Attorney General rather than a former Speaker or Lieutenant Governor to the governorship represented the changes that competition for major public office have undergone in the post-war era. In order to win a nomination for a major state office, candidates could no longer be content to acquire the support of members of Republican party organizations or even of Republican legislators. They also were compelled to appeal to a substantial section of the population which is not formally affiliated with the party organizations but which votes in Republican primaries. The Attorney General has been in a unique position to attract the attention of the press and a large segment of the electorate. Attorneys General in Iowa have launched crackdowns on liquor, gambling, and pornography immediately prior to entering a major Republican state primary. As one urban legislator commented, "The path to the governor's office is through the attorney general's office. This is how one achieves the initial necessity of getting the name known." In large measure, the shifting lines of succession in the Republican party of Iowa have reflected the transition from the predominantly rural-based support of legislative candidates for higher office to the broader electoral foundations of nominees who are recruited from other state offices. As party competition has intensified, the identification of candidates capable of capturing the interest and following of populous urban areas that are underrepresented in the legislature has become increasingly necessary. Since governors and all other major state officials must gather votes in both urban and rural areas, attorneys general have a greater opportunity to enlist the loyalty of urban voters than nominees from the legislature.

Despite the alternative route to gubernatorial power that has developed, however, governors generally have been unable to achieve the records of accomplishment that the public may expect of them. The governor has remained dependent upon the legislature for the success of his legislative program and his administration (Ross, 1957:81). "Unlike many governors, Iowa's chief executive has never assumed any great amount of legislative leadership." As one urban Republican legislator remarked cynically, "The governor has hardly any power. The legislators must really laugh to see a governor campaign and promise what he's going to do. He has less power than the lowliest legislator here."

On most occasions, the only power which the governor has been able to exercise rests on his ability to attract publicity. One Republican representative in 1963 asserted that the governor of Iowa has occupied "a ceremonial office. His influence is due solely to persuasion." A Republican state senator added that the governor "has the power to recommend, but not much more. The Interim Committee has more power between sessions than the governor."

The Interim Committee of the Iowa legislature usually has consisted of experienced members of the House and Senate who supervise the activities of the state administration between the biennial sessions of the legislature. Since rural Republicans normally have constituted a majority in the General Assembly, this group usually has predominated on the committee. In the summer of 1961, the Governor scored one of his rare victories over the Interim Committee when the committee attempted to establish a $25,000 annual ceiling on the salaries of state employees. Since the action threatened his program to employ additional psychiatrists at the state mental hospital, the Governor forced the committee to rescind its edict by appealing to the press and to the people. The governor, however, has not always been so fortunate. As one urban Republican representative concluded caustically, "The governor is the most helpless individual in state government. All he can do is recommend."

The governors of Iowa have been handicapped in their

efforts to fulfill their objectives by a number of factors. Among the most prominent limitations has been relatively brief tenure in office. Iowa governors generally have been constrained by a tradition that has enabled them "to get a second term, but nothing more" (Schlesinger, 1957:32). During the first century of Iowa history, the two-term tradition was broken only twice. Since governors of Iowa are elected for two year terms, they seldom have been able to secure their programs or to strengthen their own positions before they must face the prospect of re-election.

Iowa governors also have used the veto power sparingly. From the time Iowa became a state until 1917, for example, the governors had vetoed only 57 bills (Swisher, 1917:212). Little evidence has been acquired to indicate that modern governors have resorted to this tool more frequently. Perhaps the lack of the item veto has retarded the use of the veto power generally, but the failure to void legislative enactments also may have been associated with an unwillingness to arouse the ire of a more powerful branch of state government. Since Iowa governors usually have been beholden to the legislature for the success of their programs, as well as for their own nominations, they naturally have hesitated to alienate the legislature.

A large number of boards and commissions that administer many of the activities of Iowa government also has significantly curtailed gubernatorial authority. Since the extended terms of members generally overlap, two-term governors seldom have been able to appoint a majority of a single board or commission. The state Senate also has not been reticent in exercising a veto power through its constitutional duty to "advise and consent" to gubernatorial appointments. As a result, governors seldom have been able to effectively influence the persuasions of these administrative units.

In addition, the appointive power of the governor has been circumscribed by other factors. While relatively few public employees in Iowa have obtained their jobs through the merit system or state civil service ratings, the authority of the governor to award positions in state government has been severely limited by the independence of other state agencies and by the interference of legislators. In 1963, over 90 percent of the members of the Iowa legislature reported that they had "helped someone else get a job in state government." Apparently many of the positions in the state administration flow through the hands of the legislators rather than through the governor.

Some evidence has been collected, however, that indicates the legislature might be willing to forsake some of its power to secure effective executive leadership. In 1963, all 158 Iowa legislators were asked to evaluate the authority of the governor with regard to the legislature. In the Senate and the House, 25.9 percent of the members felt that the existing distribution of powers between the governor and the legislature was "about right." However, 51.9 percent of the legislators responded that the governor had "too little power," "not much power," or "not too much power" over the legislature. Only 8.3 percent believed that the governor "should have no more power," "should have no power," or "has too much power" over the General Assembly. While it might appear that the legislators were anxious to strengthen the position of the governor, the responses probably were more indicative of a widespread knowledge of the superior position of the legislature than of a desire to surrender legislative prerogatives.

The determination of the legislature to maintain and enhance its authority over state government perhaps has been reflected most strikingly in the scrutiny that it has exercised over state administrative agencies and departments. In 1963, all 158 legislators also were asked about the role of the legislature in the administration of state government. Only 24 percent of the members felt that the legislature should have "no more" or "less to say" about administration or that it "should not administrate." On the other hand, 39.2 percent believed that the existing relationship between the legislature and the state administration was "all right," or they found "no objection because it could be changed at any time." This attitude was summarized by a state senator who commented, "In our departmental rules committee, we make some changes to hold them down. But we all know that if things get out of hand, we can write our own ticket." Since administrative agencies have been organized primarily to carry out "legislative intent," the legislature could easily limit the authority of the state administration. Nonetheless, 27.2 percent of the legislators felt that the General Assembly should have "more to say" about the administration, or, specifically, "more to say about administrative rules."

In addition to the vigorous scrutiny that the legislature exercises over agencies and departments, the administrative authority of the governor has been restricted by the "long ballot." The various state officers, including the Secretary of State, the Secretary of Agriculture, the State Treasurer, the State Auditor, and the Attorney General have been elected by the people every two years—frequently until death or resignation has removed their names from the ballot. The election of these officials probably has reduced the power of the legislature in state departments, and it has meant that state officers have been elected by both urban and rural voters. But it also has prevented the governor from assuming principal responsibility for state programs. The customary long tenure of elected officials has inhibited the development of a coordinated policy for the state administration. As a result, the predominantly rural influence of the legislature has maintained its preeminence over the broadly based but disunited and curtailed authority of the executive branch.

Another group of beneficiaries of the "long ballot" in Iowa has been county officers. Although county officers must stand for election every two years in districts that generally correspond to legislative constituencies, they have enjoyed the supervision of a public that has been more accustomed to voting by habit than to awarding jobs to aspiring newcomers. Between 1908 and 1964, 78.3 percent of the county officials elected biennially were incumbents.

In addition, county offices in most of the 99 counties in Iowa have been the exclusive preserve of Republican candidates. Republicans occupied 77.6 percent of the county offices in Iowa during the period 1908-1964. Of the 22,526 new candidates elected to county offices between those years, 75.1 percent were Republicans. Furthermore, Republican county officeholders have been relatively unaffected by widespread popular reactions that have favored the minority party. At no election during the period 1908-1932, except in 1932, have less than 60 percent of the successful new candidates been Republicans. Even in the Roosevelt landslide of 1932, 41.2 percent of the new candidates elected to county offices were Republicans. Seemingly, Republican officials have been sufficiently ensconced in Iowa county courthouses to weather nearly any political storm that might beset them.

Furthermore, there were few major differences in rates of turnover between county offices. Only County Attorneys and Boards of Supervisors experienced somewhat higher turnover than other offices. In the elections between 1908 and 1964, 56.7 percent of attorneys and 65.3 percent of the supervisors were new candidates chosen for the two positions. The greater turnover in those offices, however, did not necessarily reflect increased party competitiveness.

On the contrary, 77.1 percent of the candidates who successfully sought the office of County Attorney were Republicans; and 70.2 percent of the victorious new candidates for the Board of Supervisors were Republicans. In no other major county office from 1908 to 1964, however, were less than 70 percent of the winning officeholders incumbents of the positions.

Although the low turnover of county officials would suggest that most county offices have not served as a source of recruitment for Republican candidates to higher political offices, the officeholders have provided the Republican party with an amazingly stable and extensive network of influence at the local level. In state government, the "long ballot" probably retarded the development of a unified executive branch that could serve as a spokesman for both urban and rural interests; but, within the counties, it has perpetuated the influence of officials who represent the same areas and who share many of the sympathies of rural-oriented state legislators. In large measure, the interests of county officials have tended to reflect rural and small town Republicanism. While the "courthouse crowd" probably has not enjoyed the power in Iowa that it has achieved in some other one-party states, the permanence and partisanship of county officials would suggest that their sway over local affairs has not been inconsequential.

Court house officials probably play a major role in local government, but they also occupy positions of considerable importance in the state legislature. The 1963 survey of the legislature revealed that the legislators visited or were visited by county officers more frequently than any other group, except the Farm Bureau. The state-wide influence of county officers, therefore, may even exceed their authority in local affairs.

Since the late 1950s, however, there has been a move in Iowa toward consolidating some county offices and placing others on an appointive rather than on an elective basis. In 1961, the first effort to consolidate two offices in Clinton County under a law passed by the legislature in 1959 provoked the opposition of County Treasurer Paul T. Eastland, secretary-treasurer of the Iowa State Association of County Officers, who argued (Des Moines *Register*, August 14, 1961) it was "a start toward centralized government." Nonetheless, there was surprisingly little opposition to the move in other quarters. In fact, during the period there was serious discussion of plans to provide officers for multi-county administrative units and to appoint rather than elect county officers.

In large measure, urban-rural configurations in Iowa politics eventually may depend upon the future of the county as a political unit. At a local level, county governments, reflecting rural interests based on geographical areas rather than population, have provided the Republican party with an extensive organization of experienced governmental officials. Similarly, in the state legislature, rural interests gained a considerable advantage from the requirement that legislative districts must be drawn along county lines rather than in terms of population. As a result, the collective interests of both the legislature and county governments have tended to underrepresent urban areas. For many years, the tendency of the Republican party to promote gubernatorial candidates from positions of legislative leadership permitted the predominantly rural interests of the legislature to be projected into the executive branch of state government.

While the recruitment of governors from other positions may reduce the impact of the legislature, it will not necessarily inspire the governor to become the major spokesman for urban aims. State elections based upon numbers rather than area provide an opportunity for populous centers to exercise a significant influence on elected officials, but they will not ensure the success of urban objectives. The limitations imposed upon the governor and the state administration probably will remain for a considerable period. As party competition increases, there may be growing incentives for the legislature to cooperate in the enactment of programs that have demonstrated their popularity in major state elections. But the eventual effects of urban-rural changes on the political life of Iowa probably will be felt only when competitiveness becomes a marked feature of state politics and when population rather than counties or other geographical units is considered the necessary basis for legislative representation.

REFERENCES

BERG, LARRY L. (1962) "Voting in the Iowa senate, 1957-1961." Unpublished paper, State University of Iowa, Iowa City.

GOLD, DAVID. (1961) "Lawyers in politics: an empirical exploration of biographical data on state legislators." Pacific Sociological Review 4 (Fall): 84-86.

HAHN, HARLAN. (1965) "Urban-rural legislative voting on taxation in Iowa." Iowa Business Digest 36 (June): 8-12.

——(1970) "Leadership perceptions and voting behavior in a one-party legislative body." Journal of Politics 30 (forthcoming).

HYNEMAN, CHARLES S. (1938) "Tenure and turnover of legislative personnel." Annals of the American Academy of Political and Social Science 195 (January): 21-31.

——and GEORGE W. CAREY. (1954) "The Iowa legislature: a general description." Unpublished paper, Indiana University, Bloomington, Indiana.

PRICE, CHARLES M. (1959) "The rural-urban conflict in Iowa." Unpublished M. A. thesis, State University of Iowa, Iowa City.

ROSS, RUSSELL. (1957) The Government and Administration of Iowa. New York: Thomas Y. Crowell Co.

SCHAFFTER, DOROTHY. (1929) The Bicameral System in Practice. Iowa City: The State Historical Society of Iowa.

SCHLESINGER, JOSEPH A. (1957) How They Became Governor. East Lansing: Michigan State University Government Research Bureau.

SWISHER, JACOB A. (1917) "The executive veto in Iowa." Iowa Journal of History and Politics 15 (April): 155-213.

WIGGINS, CHARLES W. (1965) "Party voting in the 61st Iowa General Assembly." Iowa Business Digest 36 (November): 10-18.

Chapter 

THE EMERGING PATTERN

Political conflict normally has geographical bases, but it also finds expression primarily through organized groups. In most states, political parties serve as the natural vehicles for the promotion of distinct governmental interests. Significant disputes such as urban-rural differences seldom have a major impact on state politics until they have been recognized and articulated by the parties. Urban-rural tensions probably will yield major realignments or increased political competition, therefore, only if the parties assume the role as spokesmen for the aspirations of urban and rural voters.

In Iowa, the effects of urban-rural conflict on party competitiveness has been complicated by a number of factors. Initially, insofar as either of the parties has become identified with urban interests, that group probably has tended to be the Democrats. At the same time that urban centers were gaining a majority of the population in the state, the minority party was emerging concurrently as a major repository for the affiliations of urban voters and as a principal voice for urban goals. Since the Republicans traditionally have provided the dominant and acceptable avenue for political activity, urban aims have been significantly disadvantaged by the electoral effects of the one-party legacy and by the difficulties of converting city voters to both the recent political interests of the cities and a different political party. The latter problem undoubtedly has been compounded by the relatively amorphous and flexible structure of the Republican party which historically has permitted the assimilation of new political majorities as they arose.

The tendency of parties to reflect opposing political objectives often has been mitigated through the desire of parties to avoid losing elections by alienating segments of the population or by wedding themselves irrevocably to other groups that may eventually become minorities. The ability of the Republican party in Iowa to embrace divergent and even antagonistic groups within the electorate may retard the development of urban-rural cleavages as a basis for intensified party competition. Yet, the recognition of the strength of the urban vote captured by the minority pary also might enhance urban-rural conflict and stimulate political competitiveness by demonstrating the importance of urban interests to a majority coalition. The combined result of Democratic gains in the cities and traditional Republican success in retaining the approval needed to win elections, therefore, may inspire renewed efforts by both parties to attract urban support. In addition, the intensity of urban-rural disagreements, particularly on specific and critical issues such as legislative apportionment, suggests that the development of somewhat separate sources of party strength could substantially increase political competition in the state. The examination of such propositions, however, will require some additional investigations of the characteristics of the party organizations in Iowa.

One measure of the viability of parties as well as their tendencies to reflect significant sources of political strife can be found in the party primaries. An analysis of gubernatorial primaries in Iowa from 1908 to 1964 revealed that at no time have the Republican candidates failed to secure less than 60 percent of the votes in the election. Although Iowa has a "closed primary" that requires voters to register their party affiliation, any voter can transfer to the opposite party easily by requesting another ballot and changing his registration. Consequently, the fact that over three-fifths of the voters customarily have chosen the Republican primary, even when the Democratic candidate for governor won the general election, seems impressive and significant. In addition, the maintenance of a system of party registration for primary voting has tended to perpetuate the Republican legacy in Iowa.

Significantly, the examination of the gubernatorial primaries also indicated that the number of Republican candidates seemed to bear little relation to the vote, although there was a tendency for the Republican share of the primary ballots to increase at particularly critical elections. In 1928, for example, the Republican proportion of the primary vote stood at 62.8 percent; but, by 1932, it had risen to 80.5 percent. In large measure, this trend probably reflected the proclivity of voters to seek the solutions to new problems such as the depression within the framework of their traditional party rather than to abandon their long-standing partisan loyalties. The ability of the Republican primary to retain the allegiances of most voters in the face of a Democratic landslide provided strong evidence of the success with which the majority party in Iowa has accommodated new movements and adapted to changing circumstances.

Despite the ease with which the Republican party has prevented important struggles from escaping its own ranks, however, intense clashes between commonly recognized candidates representing urban and rural interests have not been a persistent or predominant feature of Republican primaries in Iowa. At various times, candidates have emerged who appeared to bear a distinctive rural label; and, on different occasions, other Republicans who seemed to be more sympathetic to the needs of urban areas have entered races for the party nomination. But clearly identifiable urban and rural candidates seldom have waged strong campaigns against each other, nor have they occupied a prominent position in most primary contests.

The failure of urban-rural disagreements to form an enduring basis for warfare in the party primaries might be attributed in part to the growing inclination of urban voters and prospective leaders to favor the Democratic party. As Democratic candidates have represented urban objectives, competition naturally has tended to shift from the Republican primary to the general election. By the same token, the relative absence of open urban-rural conflict in the primaries also may reflect the ability of influential Republicans to discourage the development of issues that could split the party irrevocably and produce sizeable defections. While most prior factional squabbles occurred in an era when the Democrats were too weak to offer an attractive haven for potential dissidents, the emergence of urban-rural tensions has developed at a time when the Democratic party loomed as a possible major beneficiary of rifts among the Republicans. Aspiring champions of urban or rural blocs in Republican primaries, therefore, could ill afford to permit such issues to become pivotal campaign questions unless they were prepared to surrender their ambitions and their followings to the opposing party.

While urban-rural divisions seldom have been reflected in primary votes, they have been apparent in the proportions of voters who participate in general elections. In large measure, the differences also have indicated the extent to which legal provisions can influence political processes and partisan fortunes. For many years, Iowa law has required voter registration in general elections only in cities of more than 10,000 population. While the requirement apparently was designed to cope with the increased difficulty of determining voter eligibility in relatively impersonal large cities, it also has provided a sizeable advantage for small towns and farm townships and for the Republican party which generally has received greater support in those areas than in urban communities.

An examination of presidential, gubernatorial, and congressional elections between 1948 and 1958 indicates the extent to which the absence of registration requirements has inflated the proportion of the rural vote in Iowa balloting. As the bar graphs in Figure 3 reveal, there has been a consistently higher mean turnout of voters in small towns than in urban areas of more than 10,000 residents. No other dividing line approaches the importance of the level of 10,000 population in distinguishing communities with high and low turnout. Since this distinction also corresponds to a generally accepted classification of urban and rural areas, differences in rates of voter turnout clearly have a major impact on urban-rural conflict within the state.

The disproportionate turnout of voters in small towns and farming areas clearly obscured the significance of the population majority that resides in urban communities. Since urban voters participate in elections less frequently than their rural counterparts, the electoral effects of growing urbanization reached an obviously perceptible or critical stage for the parties relatively slowly. Dramatic population

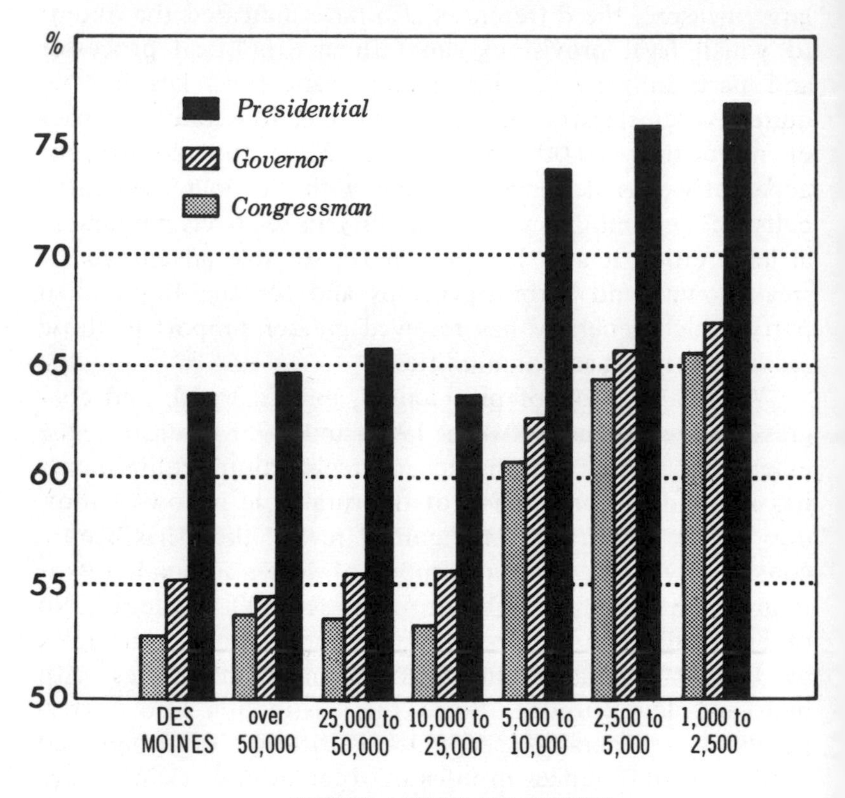

Turnout of Vote in Fifteen Elections in Iowa, 1948 - 1958
%
75
70
65
60
55
50
Presidential
Governor
Congressman
DES MOINES
over 50,000
25,000 to 50,000
10,000 to 25,000
5,000 to 10,000
2,500 to 5,000
1,000 to 2,500
POPULATION OF CITIES

Figure 3

increases in urban areas were necessary to offset the electoral advantages of rural places and to reveal the actual political strength of the cities. Until that development occurred, the higher rates of turnout in rural areas tended to reduce the potential size of the urban vote and the impact of urban-rural cleavages in Iowa elections.

The legal standards regarding voter registration not only have limited the proportion of the urban vote in state elections but they also have imposed additional duties on urban party organizations. Unlike their counterparts in rural areas, party groups in the cities must ensure that their followers satisfy the registration requirements for voter eligibility as well as attempting to persuade them to appear at the polls on election day. In some large communities, the relatively weak and disorganized nature of the party organizations has increased the difficulty of this task, particularly for the Democrats. It was not until 1956, for example, that the Democratic party launched an intensive organizational campaign in Des Moines and Polk County, the most populous urban area in the state. That year represented "the first time that every precinct was canvassed, and also it was the first time that an organized plan was used to increase Democratic registration." The results (Telford, 1957) proved that this effort was "the most successful campaign ever conducted by the Democrats in Polk County." Although the quantitative measurement of party organizational effort raises substantial problems, there is little doubt that urban party groups including the Democrats revived from a nearly dormant state as the potentially critical role of the city vote became apparent. Presumably the extension of such activities may have a continuing effect upon the political influence of urban areas in state politics. (For a study of activism in the Republican party, see Hahn, 1967.)

The evidence, however, indicates that party organizations in all areas of the state frequently have encountered major difficulties in locating the members that are necessary to maintain an effective political group. One study (Harder and Ungs, 1963) of county Republican and Democratic party committees in Iowa revealed that 98.3 percent of the vacancies in rural counties and 95.6 percent of the openings in counties containing a city of 10,000 or more population were not contested in the 1962 primary. Although both parties have experienced difficulties in finding the persons needed to staff local positions, seemingly there has been less competition for party offices in the Republican than in the Democratic party.

In 1957, a survey of 58 Republican county chairmen in Iowa (Ungs, 1957:114) found that "only 18 reported that at the time of their election there had been other candidates for the post." Similarly, in 1963 (Smith, 1963), 66.1 percent of the Republican county chairmen and vice-chairmen stated that there had been no competition for the office that they held. Only 3.4 percent indicated substantial competition for their positions. Among Democratic county chairmen and vice-chairmen in Iowa, on the other hand, only half said there was no competition for their offices, and 18.9 percent reported substantial competition.

The absence of intense competition for party offices in Iowa, particularly among Republicans, probably is related to both the relatively informal organization of local party groups and the ability of the dominant party to prevent divisive factional contests from threatening its cohesion. Most county party organizations in Iowa appear to be small and closely-knit groups in which unity is maintained by tacit agreements among the leaders. Party officeholders frequently are chosen by the officials who preceded them. In 1963, 68.6 percent of the Republican and 51.9 percent of the Democratic county chairmen and vice-chairmen were recruited by the previous party leaders. The prevalence of this system of co-option might suggest relatively rigid control of party activities by a self-perpetuating elite; but, perhaps more accurately, it reflects the willingness of party officials to accept new recruits and the relatively low prestige that normally is associated with partisan jobs. Since there usually are few potential candidates available for leadership roles, party officeholders must either draft their own replacements or remain in the position themselves.

To some extent, however, competition for party positions has become more prevalent in urban than in rural areas. In 1963, for example, 58.3 percent of the urban Democratic chairmen and vice-chairmen reported that they had faced some competition for their positions, while considerably more than half of the remaining groups of officeholders stated that they had experienced no competition whatsoever. In part, the development of growing interest in party offices in the cities may reflect the increased responsibilities and status of urban party leaders. Although urban Democrats perhaps have assumed a particularly prominent role because of the uniquely critical importance of populous areas to the fortunes of their party, the emergence of urban leaders as influential spokesmen in both political parties may be due to the large number of votes they represent and to the tasks of voter registration and canvassing they must perform. Significantly, urban leaders in both political parties also reported that they spent considerably more time working on political matters during election campaigns and at other times than their rural counterparts. Since the success or failure of parties at the polls may be more dependent on the existence of effective organizations in urban than in rural areas, urban leaders naturally may occupy more prestigious and attractive positions than rural officials.

The enhanced appeal of party leadership positions in the

cities also probably is reflected in the relatively brief tenure of urban party officials. While nearly 15 percent of the rural county chairmen and vice-chairmen in both parties in 1963 has served for more than eight years, less than four percent of the same officials in urban areas had remained for an equivalent amount of time. Almost two-thirds of the urban chairmen and vice-chairmen had occupied the offices for only one or two years. The incentives to compete for party positions in the cities may result in a higher turnover of partisan leaders in urban than in rural areas.

Urban-rural disparities in the roles of party officials also were evident in the social and economic characteristics of the men and women who held the positions. As might be expected, farmers were considerably more prevalent among rural than among urban party officeholders. Nonetheless, 12 percent of the urban Republican chairmen but none of the urban Democratic leaders were farmers. Perhaps more striking, however, was the distinction between the parties in the proportion of farmers who held party positions in rural counties. While 40 percent of the rural Democratic chairmen were farmers, the largest group of their Republican counterparts were engaged in professional or business occupations. Only 19 percent of the rural Republican chairmen were farmers. In part, such differences may have been related to the influence of the Republican party in small towns. Democrats often have scored impressive electoral gains in farm townships, but Republicans consistently have received their heaviest vote in the small communities. Apparently the separate bases of party strength in rural counties have been reflected in the recruitment of leaders.

In addition, other attributes of party officeholders appear to reveal important partisan and geographic dissimilarities. A detailed analysis (Smith, 1963) of several characteristics, for example, concluded:

> In rural counties, 15 percent of Republican chairmen are over 60 years old, compared to 31 percent in this category in urban counties. By contrast, no Democratic chairmen are over 60 in urban counties, while 21 percent are that old in rural places. Similarly ... rural Republican chairmen have higher educations and incomes than their urban counterparts, while exactly the opposite is true for Democratic chairmen.

The contrast between young, ambitious, relatively high-status Democratic leaders in urban counties and the somewhat older Republican officials in the same areas who have less income, education, or prestigious occupations than rural Republicans also probably characterizes the competition and esteem accorded Democratic leaders in urban counties.

The previous political experience of both Republican and Democratic officials also seemed to disclose significant differences between the nature of urban and rural party organizations. While 73.2 percent of the urban chairmen and chairwomen in both parties stated that they had held a prior party office, only 58.3 percent of the rural county leaders reported previous experience in a party position. No appreciable difference was found in the recruitment patterns of Republican and Democratic party groups. In both parties, urban leaders generally were promoted from lower party positions; but rural officeholders seemed to have acquired little political experience before they became chairmen or chairwomen. While partisan leadership in urban areas appeared to provide both a reward for prior service and an opportunity for further political activity, the offices in rural counties seemed to constitute a duty rather than a reward or an opportunity.

The tendency to select party leaders by cooption, with little or no competition among middle-aged or older persons from farming, business, or professional careers, strongly suggests that many officials did not regard their positions as significant avenues for political advancement. Most county chairmen and vice-chairmen in rural areas had little prior party experience and seemed to convey slight interest in subsequent political activity. For them, party officeholding appeared to be a responsibility imposed upon them by friends and associates who also had held the positions, rather than an eagerly sought prize or the object of a long-standing ambition. On the other hand, another somewhat inconsistent concept of party organizations e-merged particularly among urban Democrats. In some of the counties containing large cities, party offices were vigor-ously contested by young persons from relatively high status backgrounds who had served in party positions pre-viously. Few, however, seemed content with the titles that they achieved. For such people, county party leadership appeared to be a position they had pursued in order to launch even more extensive ambitions. Although the latter image of party office was more prevalent among Democrats than Republicans, urban groups in both parties also shared some of its important characteristics.

Urban Democrats were considerably more likely to ex-press an interest in material rewards and opportunities for personal advancement than leaders in the opposing party or in rural areas. While no ambitions for political benefits beyond their present positions were acknowledged by more than 40 percent of any of the other groups, a clear majori-ty of the urban Democratic chairmen and vice-chairmen admitted that they sought to attain personal goals. For example, 58.3 percent reported an interest in higher party office, 53.2 percent stated that they planned to run for elective office, half sought an appointive position or other sinecure, and two-thirds said that they intended to use their offices to obtain an improved private occupation. In addi-tion to their personal interests, urban Democratic officials also expressed a particular concern about the distribution of political favors and appointments to their supporters in the party organization. Unlike their Republican counterparts or party leaders in rural counties, for example, urban Democrats voiced strong agreement with the belief that effective party organizations could not be created without the aid of patronage and with the statement that a major problem for county chairmen "is the disruptive effects of patronage decisions."

The preoccupation of urban Democratic chairmen and vice-chairmen with personal ambitions as well as the material benefits of politics clearly presents some potential difficulties for the growth of a strong party system in the cities. While appointive positions may be of some value in the maintenance of established party groups, the prospective threat of disappointed job-seekers and divisive conflict over prized positions would seem to make patronage a somewhat perilous and narrow basis for the formation of local organizations. Similarly, the personal ambitions of Democratic officials might jeopardize the party in urban areas by eliminating experienced talent and by creating the possibility that the goals of the leaders and the interests of the party could conflict to the disadvantage of the organization. Since the political impact of urbanization may be influenced substantially by the development of effective party groups in the populous counties, the eventual solution to the problems raised by patronage and personal motives could have a decisive effect on the nature of party competition in Iowa.

To some extent, problems of patronage and potential factionalism are difficulties that long have plagued the Democratic party. For many years, Democrats were compelled to rely on Federal appointments rather than electoral victories as their principal sources of political satisfaction. As a result, some Iowa Democrats, like southern Republicans, began to redirect their efforts from the task of gaining success at the polls to the acquisition of appointive benefits. The shift of organizational attention not only had a destructive impact on the ability of the party to grow by winning votes, but it also tended to breed petty jealousies and squabbles between opposing aspirants for appointive positions.

Although most of the factional conflicts that beleagured the Democrats were relatively minor local affairs revolving around personalities, they probably undercut the effectiveness of the party organization by focusing energies on internal disputes rather than on the task of gaining external popular support. In part, the growth of the party, particularly in urban areas, may obviate such trials by demonstrating the opportunities for electoral success; but the legacy of defeat and despair may continue to plague the Democrats for many years.

During a substantial portion of the era after World War II, the Democratic party in Iowa was controlled largely by the state chairman, Jake More. At the time, there was considerable disagreement "as to whether the Democratic organization should be called a 'machine,' but most people believe that it has too many individualists and political mavericks to make machine politics possible in a state dominated by the other political party" (Mashek and Johnson, 1954:208). Many observers felt that More was deposed in 1958 at the instigation of Governor Herschel Loveless. Opinions of More have continued to vary, and his followers retained an active role in the party. But many Democrats might have echoed the sentiments of one urban precinct worker who said, "Jake More was like a southern Republican politician. He had no interest in winning."

In 1960, the leadership of the state Democratic party was transferred to Lex Hawkins, a Des Moines attorney, who had been elected chairman of the Polk County organization at the age of 26. While Hawkins too had his detractors as well as his supporters, it probably would be conceded that he brought a new, primarily urban image to the party in a period when its prospects for success were considerably brighter than they had been a few years earlier.

Although many of the problems of the Democratic party probably were overcome by the growing likelihood of victory at the polls, a third major difficulty has remained. Since the Democrats traditionally offered few opportunities for political advancement in Iowa, they frequently have been unable to locate men who would be attractive nominees for public office. As one former Democratic Congressman confessed, "The party has had trouble sometimes finding local candidates who a person wouldn't be ashamed to stand on the same platform with." The problem even has extended to higher political offices. In 1958, when four Democratic Congressmen were swept into office in Iowa, three subsequently faced nepotism charges for employing relatives on congressional payrolls. One of the Congressmen, Merwin Coad, also became involved in financial and divorce scandals that may have done considerable damage to the prestige of the Democratic party.

In large measure, the responsibility for recruiting both Democratic and Republican candidates normally has been shouldered by local party organizations. One of the common preoccupations of chairmen in both parties has been the task of finding a sufficient number of local candidates to "fill the ticket." Naturally, this pursuit has been more difficult for Democrats than for Republicans, particularly in the rural counties where Democratic nominees have few prospects for success. In 1944 (Porter, 1945:733-734), for example, the Democrats nominated only 283 candidates for the 891 county offices which they were entitled to seek. "In 34 counties, there were no Democratic candidates for nomination for any county office." Although electoral gains and improved organizations have resulted in a subsequent increase in the number of choices available to voters, the "long ballot" in Iowa has forced the parties to devote a considerable amount of time and energy to the identification of persons who would serve as candidates for county offices.

Political traditions in Iowa generally have made candidacy for public office dependent upon party encouragement rather than independent volition. In 1963, for example, 69 percent of the members of the legislature claimed that seeking office was not their own idea but that they had been encouraged to run by someone else. Of the members who said that they were persuaded to become candidates, 44.9 percent, or nearly one-third of all legislators, cited their party organizations as the principal sources of encouragement. Only 11.9 percent reported that former legislators sought to influence the choice of their successors, and the remainder asserted that they were encouraged by friends, neighbors, or associates in other activities.

Local party organizations seem to play a prominent role in the selection of nominees, even among candidates who were recruited primarily by other groups. A rural Republican representative reported, for example, "I was talked into it by Farm Bureau people. Of course, they also checked with the chairman of the county central committee." On some occasions, the efforts of party groups to obtain an acceptable candidate may extend to the promotion of primary contests. Another rural Republican commented, "The committee didn't want the man who announced his candidacy. They encouraged me." Informal caucuses and party committees frequently may have had a major influence on the choice of candidates for many years, but no official system of pre-primary endorsements existed in Iowa until 1964 when the Democrats adopted it. While the award of party sponsorship to candidates before the primary election might stimulate increased loyalty to party policies and programs in urban areas where the Democratic emblem could be electorally significant, it may be no more effective in rural counties than traditional party attempts to recruit candidates.

Some evidence also suggests that the concept of party responsibility has been more firmly embedded in urban than in rural areas. The survey of county chairmen and vice-chairmen, for example, revealed that urban leaders in both political parties were considerably more likely than their rural counterparts to frequently let elected officials "know what [the party] would like them to do." Similarly, the statements of both Republican and Democratic urban members of the legislature indicated that they had more numerous contacts with local party officials than rural representatives. Since urban party organizations often have had distinct interests to protect and promote, they apparently have developed a closer working relationship with public officials than partisan groups in rural counties.

A strong sense of party allegiance probably can not be established among public officeholders, however, until the parties are capable of playing a critical role in the election of candidates. Although local party groups in Iowa customarily have performed an important function in encouraging candidates, they seldom have been able to provide more than psychological assistance and comfort. When all the members of the Sixtieth General Assembly were asked if their party had given them any support, 73.4 percent responded affirmatively. Only 8.2 percent replied that the party had given them "some support," and 18.4 percent denied that they had received any campaign aid from party organizations.

Yet, the vast majority of the legislators also made it clear that they had obtained little, if any, material assistance from the party. Most of the representatives from both political parties reported that the main sources of party help consisted of "talking up" the legislator's candidacy and an occasional newspaper ad calling on voters to support the entire party slate. Characteristic was the reply of one rural Republican who said, "Yes, I got good support from the party. No financial support, but terrific moral support."

While Republican strength in many areas of the state has made extensive campaigning needless, Democratic candidates often have not received effective backing because of the weakness of local party organizations. As one Democratic senator from a district containing a moderately large city complained: "I was given general support, but very little financial help. We always have a problem of party responsibility and loyalty. One of the best ways of gaining this loyalty is through financial support. But, lacking this, the candidate is not bound." The nature of partisan competition in Iowa has not provided either political party with strong incentives to offer its candidates tangible campaign assistance. As a result, the parties generally have been unwilling or unable to exercise effective discipline over elected public officials, despite their active involvement in the recruitment of candidates.

The energy expended by the party in recruiting candidates as well as the relatively extensive political or civic experience of most legislators indicates that many Republican nominations for major local offices frequently have been regarded as a prize for outstanding party members rather than as an opportunity for further political advancement. Since Republican county candidates seldom have extensive governmental aspirations, there have been few effective means by which party groups can control their activities after their election. Similarly, the fact that Republican nomination usually has been tantamount to election, especially in many rural counties, has obviated the necessity for extensive campaign assistance. On the other hand, Democratic candidates in similar localities often have been characterized by a sufficiently intense sense of party loyalty to inspire them to serve in a sacrifical capacity on many occasions. Since such nominees normally have been unsuccessful, however, their adherence to the party has had little impact on the political characteristics of the state.

The problem of obtaining the necessary resources for a local campaign, however, has formed a major obstacle for candidates in both parties in Iowa. In 1962, for example, only 36.8 percent of the Republican and 54.5 percent of the Democratic nominees for the Iowa legislature received any campaign contributions whatsoever. The largest group of candidates, therefore, were compelled to assume the costs of campaigning as a personal expense.

The slight financial advantage enjoyed by Democratic contenders probably was related not only to their increased difficulties of obtaining voter recognition and familiarity but also to the support that they received from their party. While the bulk of the Republican contributions were acquired from volunteer committees and individuals, the Democrats were more successful in soliciting donations from labor unions and party committees. Among the candidates for the upper house of the legislature, for example, 20.3 percent of total contributions received by Democratic candidates were donated by party committees. By contrast, only 9.1 percent of the Republican money came from party sources. Republicans also allocated larger proportions of their campaign budgets for donations to party coffers than the Democrats. In some counties, such contributions apparently have been placed on a regular basis. Two Republican candidates in 1962, for example, listed their party donations as "assessments" in the official reports. While Republican candidates for the state senate reported $295 in donations to their party as "campaign expense," they received only $565 from the party organization. Democratic nominees for the same office, on the other hand, were given $810 by their party without having made any contributions to the party treasury themselves. The relative lack of financial support for Republican candidates from the party and from other sources probably has reflected both the success of the party, which often has rendered extensive campaign exertions unnecessary, and the status of nominations by the dominant party for county offices. Since Republican sponsorship customarily has been viewed as a reward rather than a threshold for increased political activity, there has been little need for the party to bestow favors in addition to the nomination or for the candidates to wage intensive campaigns.

The state-wide campaigns of both parties have reflected the same limited contributions and expenditures that usually characterize political activities at the local level in Iowa. Although both the Republicans and Democrats maintain relatively exclusive "clubs" in which membership is based on substantial financial gifts, generally there have been few prominent party "angels," or large contributors, in either party in Iowa. In the Republican party, a considerable amount of money has been collected by the Republican Finance Director, who frequently has been promoted to the state chairmanship of the party. The position was created in 1948, as a former Republican state chairman emphasized, "so that the names of the contributors would not be in front of the candidates or their assistants."

The relatively slight importance attached to financial contributions by most Iowa politicians perhaps has resulted from the lack of extensive needs for campaign resources. The estimates of most Iowa Congressmen, for example, revealed that their campaign costs seldom have exceeded $20,000. Since this amount probably would be considered low in comparison with campaign budgets in other states, candidates in Iowa have not faced the financial hardships that have afflicted nominees for public office in other regions of the country. As a United States Senator, who reported that the costs of his first campaign did not exceed $5,000 in the primary and $50,000 in the general election, commented, "Politics is conducted on a relatively austere basis in Iowa."

In large measure, interest groups frequently have assumed a major role in providing financial assistance for candidates in Republican primary contests, where elections often have been decided in Iowa. As a perceptive rural Republican legislator acknowledged candidly:

> I think the powerful lobbyists like . . . can raise a lot of money in a primary contest. But it's all done over the telephone, and it's almost impossible to prove. They'd deny it if you asked them.

On occasions when the principal issues in primary contests have revolved about urban-rural considerations, the financial support of interest groups also has tended to divide along similar lines. A former state Republican leader admitted, for example, that in the 1948 gubernatorial primary the incumbent Robert Blue was "gctting strong support financially from the Iowa Manufacturers Association," while his opponent William Beardsley was receiving equally large contributions from "Farm Bureau members to enhance Beardsley's coffers." As in the general elections, however, campaigning for primary nominations in Iowa seldom has formed a very elaborate or expensive operation.

The purposes for which contributions were spent also revealed some differences betwen the parties and some distinctions between campaigning in urban and rural constituencies. An analysis of campaign expenditures by candidates for the state legislature in 1962, as an example, suggested that Republicans may have engaged in a more personal style of campaigning than Democratic nominees. Most Republicans allocated a relatively small portion of their campaign budgets to advertising, but they spent significantly more than their Democratic counterparts in an effort to reach the voters directly through the mail. The chief financial beneficiaries of campaign expenditures by candidates of both parties, however, were the newspapers. Slightly less than one-third of the money spent by Republican and Democratic nominees for the legislature was devoted to notices in local newspapers. The large number of daily and weekly newspapers in Iowa probably have performed a critical role not only in disseminating information about political events but also as an important means by which candidates could communicate with the public.

In addition, campaign expenditures for the legislature also revealed substantial differences betwen urban and rural candidates. As might be expected, legislative aspirants of both parties in urban areas spent considerably more money on advertising through mass media such as radio, television, and billboards than their rural counterparts. Although most campaigns in Iowa have not reflected major promotional efforts, five urban candidates for the upper house of the state legislature paid the bulk of their campaign expenses to advertising agencies. Since most of the mass media as well as the advertising firms are based in urban centers, the increased costs of such methods of appealing to the voters may affect styles of campaigning as urbanization increases.

The introduction of radio, television, and other means of mass communication, however, apparently has had little initial impact on electoral patterns in Iowa. One study (Simon and Stern, 1955:475), which compared counties served by television stations with counties that had not yet received television coverage in the 1952 presidential election in Iowa, found that there was no "basis for asserting that television had a significant effect upon voting participation or upon the party division of the vote." While television and other media may produce relatively slight effects on voter decisions, they may inspire significant changes in Iowa politics as increasingly important means by which candidates can achieve voter familiarity and exposure.

As a continuously important political force in Iowa, however, newspapers probably outrival any other form of communication. In the latter part of the nineteenth century, the influence of "Ret" Clarkson, editor of the *Iowa State Register*, in the Republican councils of the state was so great the period often has been referred to (Cole, 1921:396-397) as the "Register Regency." Similarly, the persuasive appeal of the farm publications operated by the Wallace family has been strongly felt in political circles in Iowa. In 1887, for example, "Uncle Henry" Wallace plunged his publication, *The Homestead*, into state legislative politics and boasted that he "managed to defeat six Republicans and six Democrats" (Lord, 1947:103). The impact of the press has not always been as dramatic or as bipartisan as the Wallace effort, but the editorials and reports of political events in Iowa papers undoubtedly have played a prominent role in shaping political attitudes and partisan preferences.

The influence of newspapers in Iowa politics has not only been exercised by powerful major journalists, but it also has been wielded by the editors of the daily and weekly papers that exist in nearly every county in the state. The small town weeklies seldom have been active in the promotion of urban interests or objectives. The only major medium of communication, other than some radio and television stations, encompassing the state with a generally urban viewpoint on critical issues such as legislative apportionment, is the Des Moines *Register*, which also has exhibited moderate Republican leanings. Except for this paper, which perhaps has provided the major record of state political activities, most journals have reflected a primarily rural vantage point. In addition, most local newspapers in Iowa usually have been Republican. Although the effects of an overwhelmingly Republican press have been difficult to isolate precisely, they probably have contributed to the maintenance of traditional party loyalties and to the precedence of the party in state politics.

Republican preeminence also has been fostered by common devices that normally have been employed by political parties to ensure political control in a state. In particular, the art of the gerrymander, or the drawing of legislative district boundaries to maximize opportunities for the majority party and disadvantages for the minority party, frequently has been practiced in Iowa. The legislative acts for congressional redistricting in 1882 and 1886 provided, in the estimation of one commentator (Rystrom, 1961), "an outstanding example of what can be done toward weighting districts toward the party in power." By combining a major area of Democratic strength into one district and by attaching other Democratic counties to large Republican areas, the Republican legislature not only retained a congressional majority at that time but also "in later years it helped to keep Democratic congressmen to a minimum."

Subsequently the technique of designing political districts to fulfill the interests of the majority party has continued. In 1961, for example, a Republican legislature refashioned the state into seven congressional districts by linking two urban Democratic counties into one chair-shaped district. As a Republican legislator admitted candidly, "I don't think we want to be fair and impartial here. What we want is to set up a plan to elect as many Republican congressmen as we can." Since the establishment of congressional boundaries involved a basically political decision, however, the legislator hardly could have been censured for his honest evaluation of partisan considerations.

The apportionment of congressional districts clearly may have a major effect upon the representation of urban areas in the federal government. Since large towns in Iowa are widely dispersed throughout the state, the interests of urban voters might be muted somewhat by the merger of urban and rural counties in single congressional districts. On some occasions, however, as in 1961, the natural impact of geographical factors on congressional districting may be modified by concerns of the parties. While the topography of the state normally would yield little primary representation of urban centers in Congress, such factors can be overridden by the partisan characteristics associated with urban and rural areas.

Perhaps more important than congressional districting, however, is the apportionment of state legislative constituencies. Since state legislators represent smaller and more homogeneous areas than Congressmen, the geographical distribution of the population seldom can allay the divisive tendencies of urban-rural differences in legislative districts. While most state institutions do not contain distinct urban or rural representatives and major executive officials are selected by voters in both areas, the legislature provides a common arena for conflicts between members who are closely linked to the characteristics of their constituencies. As a result, urban-rural disagreements have been projected in the legislative councils of states perhaps more frequently than in any other political body.

In addition, disputes concerning state legislative apportionment traditionally have not been confined to the location of boundaries but they have focused on the extent to which population should be considered in the creation of legislative districts. While broad federal guidelines generally have preserved at least a measure of equality in the size of congressional districts, substantial population disparities have developed at the legislative level. Contests regarding legislative apportionment, therefore, normally have resulted in direct clashes between urban and rural residents rather than between the parties or interest groups. In fact, urban-rural conflict perhaps has been more evident in debates on reapportionment than in any other controversy or issue.

The intensity of urban-rural divisions on the apportionment question probably is understandable in the context of state government. As previous analyses have indicated, perhaps no other body occupies a more crucial role in state politics than the legislature. Not only do legislators determine policies that may benefit or harm urban and rural interests, but they also play a prominent role in the political parties, the activities of lobbyists, and the selection of candidates for higher public office. The representation of urban and rural areas, consequently, can have an effect that permeates the political fabric of the state. Since the battle over legislative reapportionment has yielded perhaps the clearest expression of urban-rural conflict in Iowa politics, the portentious implications of this controversy are worthy of careful attention and scrutiny.

Although the Iowa Constitution of 1857 required the reapportionment of the state legislature according to population every ten years and a major reapportionment plan had not been enacted since the beginning of the twentieth century, there was no major effort to secure a change in the Iowa legislature for nearly fifty years. By the 1950s, however, urban interests in the state had become sufficiently distinct and pressing to stimulate renewed efforts to redress the over-representation of rural areas in the legislature. After the defeat of a series of reapportionment plans, discussions by frustrated urban Democratic legislators with Robert Johnson, administrative assistant to Governor Herschel Loveless, resulted in the formation of a bipartisan organization to promote legislative reapportionment (Schmidhauser, 1963).

The principal technique that the group, known as the Citizens Committee for a Constitutional Convention, relied upon to achieve its objectives was another state constitutional requirement which necessitated a decennial vote on the question of holding a state convention to revise or substitute the constitution. The group sought to persuade voters to support a constitutional convention in the 1960 election in the hope that the convention would adopt new apportionment provisions which would give increased representation to urban areas in the state legislature. The Committee was joined by the state AFL-CIO, most of the urban press, the Iowa League of Women Voters, the Iowa Farmers Union, and the Democratic party in endorsing the constitutional convention to secure what they considered an equitable apportionment system. Conspicuous by its absence in this coalition was the Iowa League of Municipalities, which enjoyed exceptionally good relations with the legislature. As the League explained its neutrality in the controversy, "Municipalities are dependent on the good wishes and cooperation of the state legislature for their survival." Ironically, the two strongest opponents of the constitutional convention were the Iowa Manufacturers Association and the Iowa Farm Bureau, organizations which perhaps maintained the most influential lobbies in the legislature. The battle was fought primarily, therefore, between groups which had relatively little to fear from the reprisal of rural legislators and interests which had established particularly influential relations with the legislature.

The alliances formed in the 1960 campaign for a constitutional convention also reflected the natural culmination of enduring conflicts in Iowa politics. Most of the groups that favored the convention, including labor, the major city newspapers, and the Democratic party, also had become identified with urban goals; but their influence among legislators hardly would have been classified as predominant. On the other hand, the interests opposing constitutional change, including most small newspapers, business, and the Farm Bureau, were extremely powerful in the rural-oriented legislature. The clash between the respective groups was not confined to the constitution or to reapportionment, but it had been a prominent feature of debates concerning most major governmental policies and programs in Iowa.

The controversy stimulated by the reapportionment movement produced the most direct confrontation between the principal antagonists in Iowa politics perhaps because it threatened to change drastically the strength of the contending groups. A redistribution of urban and rural representation not only would alter the composition of the legislature but it also might significantly affect the ability of the opposing sides to obtain their policy objectives and aspirations. As a result, differences in the appropriate weight to be given urban and rural interests probably were at the heart of disagreements concerning most other significant plans and proposals in the legislature.

In addition to the prominent role played by the interest groups in the 1960 campaign, the positions taken by the political parties reflected an interesting feature of Iowa politics. Significantly, while the Democrats explicitly endorsed the constitutional convention and reapportionment in their party platform, the Republicans tended to assume a relatively neutral or ambivalent stand on the issue. In part, this posture may have been related to the traditional efforts of Republican leaders to accommodate divergent and conflicting groups within their party. But the statements of the parties also clearly demonstrated that the Democrats were willing to cast their lot almost exclusively with urban interests on this issue, while the Republicans still hoped to maintain the support of both urban and rural voters.

Although the outcome of the referendum seemed to provide a clear victory for rural interests, the accuracy of the vote as a reflection of public opinion or as an indication of the strength of urban and rural forces was obscured by the mechanics of the election. The constitutional convention was defeated by a vote of 534,628 to 470,257, but city voters faced a peculiar handicap in the referendum. While residents of urban counties cast their ballots on voting machines, most rural voters marked paper ballots.

The difference between the two methods of voting, while seemingly trivial, could be of particular importance to constitutional referenda, which frequently are overlooked by the electorate. A prior study (Mather, 1960:33) of the vote on special questions in Iowa had found:

> There is a clear tendency for fewer votes to be cast on special questions when the questions are voted on voting machines rather than on paper ballots. . . . When voters in voting machine counties are given paper ballots on which to vote on special questions, their level of participation is much closer to that of the voters in the regular paper ballot counties.

The study concluded prophetically that "urban counties would stand to lose a substantial portion of their voting strength" if voting machines were used "on questions involving an urban-rural division." Therefore, like the requirement that residents of cities of more than 10,000 population must register to vote, the use of voting machines on special questions probably limited the electoral power of urban areas.

The results of the 1960 referendum on a constitutional convention of Iowa clearly revealed the differences in voting practices:

> Sixty-four of the rural counties utilized a paper ballot on the issue. Ninety per cent of the electors in these counties who voted for President also voted on the convention issue. By contrast, nine of eighteen urban counties voted on machines. In these counties only 59 per cent of the electors voting for President participated in the convention vote. The differences in levels of participation could well have tipped the scales in this test of urban-rural political power [Schmidhauser, 1963:29].

While the neglect of the convention question in the presidential election by many city voters may have reflected a lack of consciousness of distinct urban aspirations, it also illustrated the institutional barriers that have retarded the promotion of urban goals.

There was little doubt, however, that the vote on the constitutional convention was highly related to urban-rural differences. Two studies, for example, disclosed a close association between the vote by county on the convention issue and indices of urbanization in Iowa. One researcher (Mather, 1960) found a correlation coefficient of +.89 between the two variables, and another (Wiggins, 1963) discovered a correlation of +.62. Although the figures varied because of slightly different definitions of "urbanization," it was clearly evident that the urban-rural cleavage played a prominent role in the constitutional referendum.

The vote on the convention also revealed that the urban-rural conflict apparently tended to overshadow partisan differences at least in urban areas. Although the Democrats constituted the only party that explicitly endorsed the convention to secure reapportionment, urban Republicans were nearly as favorable to the issue as Democrats in the cities. In general, "the top Republican precincts in cities with a population of 80,000 or above were heavily for a constitutional convention, although generally not quite as strongly as the top Democratic city precincts" (Schmidhauser, 1963:30). Support for the issue in urban areas was not differentiated by socioeconomic levels or by partisanship. Another study in Des Moines (Salisbury and Black, 1963:591), for example, found no statistically significant relationship between the vote on the 1960 convention question and social class or party preference. "Both party groups and all classes tended to divide in similar ways, favoring the convention by a margin of four to one." Despite their partial neglect of the issue, therefore, most urban voters regardless of their social or partisan positions were strongly united in behalf of a constitutional convention to obtain increased representation in the legislature.

The outcome of the referendum indicated, however, that the relative cohesion of the urban vote was insufficient to offset the large number of ballots that were cast in opposition to the convention in rural areas. While this result clearly reflected the organizational power of rural groups such as the Farm Bureau, it soon became apparent that at least some form of reapportionment in the Iowa legislature was nearly inescapable. As a consequence, the Farm Bureau sought to preserve its victory by introducing an apportionment plan of its own which would reflect some concessions to urban areas but which also would ensure the maintenance of a rural majority in the legislature.

The legislative sponsor of this proposal in the 1961 session of the legislature was Republican Senator David Shaff from predominantly urban Clinton County, who claimed that he personally would have preferred another system of apportionment but felt that the Farm Bureau measure was the only plan the legislature would accept (Des Moines *Register*, February 28, 1961). The so-called Shaff plan provided for a House of Representatives in which representation was based on area, with one representative from each county, and a Senate apportioned on population with districts which could not vary by more than 10 percent in size. As a result, the plan would have reversed the normal system by making the Senate the popular chamber and the House the malapportioned body, and it would have established the county unit as a basis for area representation. At the same time, the plan would have reduced the size of the House from 108 members to 99, while adding only eight population representatives in the Senate. The constitutional requirement that any amendment to the Iowa constitution be passed by two successive sessions of the legislature as well as by a popular vote of the people meant that the Shaff plan occupied center stage in the controversy over reapportionment during the critical period from 1961 to 1963. While subsequent action by the courts made the debate a partial exercise in futility, the discussions of the Shaff plan revealed the broad implications of urban-rural cleavages in Iowa politics. Since the battle over reapportionment revolved about the Shaff plan more than any other proposal, the plan deserves perhaps even more attention than the reapportionment measure that was finally adopted by the legislature.

The Shaff plan probably confronted its most determined opposition in the state Senate in 1961. Unlike the state-wide party organization which had assumed a relatively neutral position on the issue, however, Republican leaders in the legislature joined the Farm Bureau in exerting strenuous efforts to secure the adoption of the Shaff plan. One rural Senator who had opposed the plan in 1961, for example, charged that he had been offered an important committee assignment in exchange for his support of the plan and that the votes of other Republican Senators were changed by appointments to committees, state administrative boards, and judicial positions. While definitive proof of such allegations may have been lacking, there was considerable evidence to indicate that a number of favors and rewards were distributed in the Senate in an effort to influence votes on the Shaff plan.

Democrats, on the other hand, remained relatively united in their opposition to the proposal. Only one rural Democratic Senator broke ranks to vote for the plan, largely as a result of pressure from the Farm Bureau. In a nearly unprecedented move, the Democratic State Central Committee officially censured the maverick senator, Adolph Elvers of Elkader, and barred him from future party caucuses (Des Moines *Register*, March 7, 1961). While the censure may have deprived the senator of his party identity, it apparently did not materially damage his relations with his constituency. In fact, a prominent state Farm Bureau official admitted that a Bureau-sponsored candidate who was anxious to oppose Senator Elvers subsequently was discouraged by the Bureau from entering the race against Elvers. Nonetheless, the action by Democratic party organization not only reflected the extreme commitment of the Democrats to the cause of urban voters but it also exemplified the membership discipline that can be imposed by a minority party which does not face the task of appealing to the separate and disparate groups that frequently co-exist within a majority party coalition.

The Shaff plan passed the Iowa Senate in 1961 only after Lieutenant Governor William Mooty broke a 25-25 tie on a critical vote by exerting his influence in behalf of the proposal. Although the plan still required approval in the 1963 session of the legislature and in a popular vote, a number of events occurred in the meantime that seemed to imply contradictory probabilities regarding its eventual success.

On March 26, 1962, the United States Supreme Court

in the case of *Baker v. Carr* concluded that the issue of the apportionment of state legislatures was justifiable under the "equal protection of the laws" clause of the Fourteenth Amendment to the U.S. Constitution. Although the court initially declined to specify satisfactory standards of population equality, the decision that courts would examine the issue of the apportionment of legislative districts raised fresh doubts regarding the legality of the neglect of state constitutional provisions on reapportionment and of the Shaff plan. Almost immediately, the President and Secretary-Treasurer of the Iowa Federation of Labor, AFL-CIO, filed suit in federal court to challenge the constitutionality of the existing apportionment of the Iowa legislature as well as the Shaff plan. In this litigation, the entire question of apportionment was examined thoroughly by the courts (Hahn, 1963).

In defending the current apportionment of the legislature and the Shaff plan, attorneys for the state placed considerable reliance on the failure of Iowa voters to approve a constitutional convention for reapportionment as well as on "the legislator's function as a liaison between the county and state government." The court dismissed the first contention by noting that the voters "have not ratified the present apportionment system." The second argument, on the other hand, was particularly important because of the use of the county as the principal basis of representation in the House of Representatives under the Shaff plan. Yet, the requirements of inter-governmental relations did not seem to form a sufficient justification for an apportionment based on counties rather than on population. Historically, in Iowa, as the court declared, "Population has been a central basis for apportionment of the General Assembly from the time of its territorial predecessors." While the grant of a legislative majority to counties was not sustained by the courts, examinations also were conducted on the influence of other segments of the population in the legislature under different apportionment schemes.

The assumption often has been made that unequal apportionment of legislative seats may give dominant representation to geographical areas or to farmers. Yet, as the brief for the plaintiffs demonstrated, the 55 least populous counties, which formerly controlled a majority under the old apportionment of the lower house of the Iowa legislature, reflected only 51.8 percent of the area and 46.5 percent of the farm residents in the state. The 50 least populous counties, which would have constituted a majority in the Iowa House under the Shaff plan, represented only 45.5 percent of the territory and 40.9 percent of the farmers in Iowa.

Perhaps even more striking was the fact that farm residents did not form a majority within the 50 or 55 least populous counties of the state. Statistics from the 1960 census revealed that farm inhabitants constituted only 40.7 percent of the people living in the 50 least populous counties and 40.6 percent of the population of the 55 least populous counties.

The Shaff plan and the existing apportionment system provided small town voters rather than farmers or urban dwellers with majority representation in the Iowa legislature. Persons living in small incorporated towns of less than 10,000 accounted for 52.1 percent of the population of the 50 least populous counties and 52.6 percent of the people in the 55 least populous counties. Yet, the residents of the 20 counties containing cities larger than 10,000 represented 53.1 percent of the total population of the state. Within the 20 counties, the people living in communities of 10,000 or more population composed 71.8 percent of the inhabitants of the counties. As a result, voters in cities of more than 10,000 population probably were the most underrepresented bloc in the Iowa legislature, while the residents of towns smaller than 10,000 were perhaps the most overrepresented group.

There also was some evidence that the underrepresentation of the cities in the legislature had impeded the passage of urban policies and programs. The opinion of the court in *Davis v. Synhorst* (217 F. Supp. 492, 1963) observed:

> There is evidence, as might be expected, of the existence in the Legislature of economic blocks, such as the farm block, the insurance block and the cities group, and that blocks having the largest membership are often more successful in their legislative programs.

In addition, plaintiff Charles L. Davis, state president of the Iowa Federation of Labor testified that populous counties were allocated fewer funds by the state than they pay in state taxes because urban representatives in the legislature "haven't got the votes to do anything about it." Furthermore, as prior analyses have demonstrated, substantial differences were evident in the votes cast by urban and rural legislators not only on taxation measures but also on legislation concerning economic or business regulation.

Despite the fact that the rural or small town segment of the population was "substantially overrepresented in the Iowa legislature and that this disparity probably is reflected in legislative voting," the court refused to rule on the constitutionality of the Shaff plan until action on the proposed amendment had been completed. Although many felt that the opinion cast considerable doubt on the constitutional acceptability of the Shaff plan, the court confined its holding to the system of apportionment that had existed in Iowa for nearly fifty years prior to the introduction of the Shaff plan. On that matter, the court concluded that "the disparities present in the apportionment of both houses of the Iowa General Assembly transgress the constitutional limits of equal protection."

While judicial activities confirmed the necessity of a reapportionment of the Iowa legislature based on population, there were growing indications that the Shaff plan was likely to be adopted by the state legislature. By 1963, the Republican party generally had embraced the proposal, and its only remaining opponents consisted largely of Democratic senator observed, "The Farm Bureau had this thing passed its second session of the Iowa General Assembly with little major difficulty. As one disgruntled urban Democratic Senator observed, "The Farm Bureau had this thing all greased. They had their 30 votes, and because of the fact that the Republicans endorsed it, this held some votes that the Farm Bureau didn't have." The final barrier to the approval of the Shaff plan as a constitutional amendment, therefore, was a state-wide referendum.

The Shaff plan referendum was held on December 3, 1963, when the ballot did not contain other contests or questions that could have affected the vote on reapportionment. Although the Farm Bureau again exerted intense organizational and publicity efforts on behalf of the proposal, the voting strength of urban areas was dramatically revealed in this election. The Shaff plan was defeated by a margin of nearly 60 percent in the state. The proposition received a majority of the votes in 64 of the 99 counties, but overwhelming opposition was recorded in 17 of the counties containing cities of more than 10,000 population. Early returns, for example, revealed that there were 271,214 votes in opposition, with 163,417 coming from the 17 counties. The 'yes' vote was 191,421, with 48,113 being cast in the urban centers.

The question also apparently drew an increase in voter participation in urban areas. The state "voter turnout of more than 466,000 exceeded expectations and was substantially greater than the total state-wide primary vote in either 1960 or 1962" (Des Moines *Register*, December 4, 1963). However, in rural Greene county, for example, the

Shaff plan attracted 757 fewer voters than a county referendum on liquor held nine days later, and 1,622 fewer voters than the 1962 gubernatorial election.

The holding of the referendum at a time when urban-rural conflict and agitation over reapportionment was perhaps at a peak seemed to produce some unusual cleavages. Although the Shaff plan would have given majority representation to small towns rather than farm areas, farm townships in 19 counties supported the plan with a 70.2 percent margin. On the other hand, the county seat towns in the 19 counties opposed the plan by 56.4 percent (Hahn, 1964). Small town voters apparently identified with the residents of big cities on the issue of reapportionment, while farm electors favored the Shaff plan, even though the latter group probably would have received fewer advantages from the proposal than small town inhabitants. Unlike other political issues and elections, therefore, reapportionment apparently stimulated at least a partial merger of urban and small-town voters in opposition to farm residents.

In large measure, the peculiar political alignments evident in the vote on the Shaff plan may have been inspired by the tendency of many small towns to regard themselves as potential metropolises and by the intensity of urban-rural conflict over reapportionment. Since the publicity generated by the Shaff plan seemed to pit cities and towns, regardless of size, against rural or farm interests, voters apparently responded to the issue without carefully evaluating the benefits and disadvantages of the plan for different segments of the population. While continued small town support for urban interests may be a dubious possibility, the urban-rural split that was sharpened by the reapportionment controversy may become an enduring schism in Iowa politics.

Since the electoral defeat of the Shaff plan left Iowa with only the existing system of apportionment that already had been found constitutionally objectionable, the courts were compelled to ask the legislature to adopt a new apportionment scheme. After a fierce battle in a special session of the Sixtieth General Assembly, a new apportionment plan finally was adopted for the 1964 elections that substantially increased the representation of urban areas in both the House and the Senate. Although the details of the new plan as well as other reapportionment proposals continued to generate controversy, legislative action as well as judicial standards had firmly established perhaps the most important principle in the conflict by ensuring that equality of population will form the major basis for the apportionment of legislative seats.

While the guarantee of population equality in legislative representation might reduce the strength of urban forces by removing an important objective from political contention, there is an equal likelihood that the reapportionment victory could provide the basis for growing political competition in Iowa. The grant of increased legislative representation to urban areas may result in a greater likelihood that not only will urban legislators be able to achieve or at least to compete on a relatively even basis for major state policies and programs but also that the cities will be provided with enhanced opportunities to recruit major state candidates and to develop effective party organizations. Since the legislature occupies a pivotal role in the political system in Iowa, the subsidiary benefits available to the cities from increased legislative representation could produce major political changes by encouraging the leaders of both parties to compete vigorously for the urban vote and by permitting urban voters to participate directly in the formulation of state policy.

Considerable disagreement, however, has existed regarding the impact of growing urbanization on party competition, apart from specific issues, organizations, or political cultures. While the general belief has been that the heterogeneity and numerical importance of urban areas usually provoked increased party competition, one study of partisan successes in local elections in Iowa from 1946 to 1956 (Gold and Schmidhauser, 1960:74) concluded that "among the 99 counties of the state there was *not* a simple positive association between the degree of urbanization and the intensity of party competition." In part, the findings may have been affected by peculiar features of Iowa politics during that period, such as the relative scarcity of two-party systems in most counties, "Democratic party weakness," as well as the definition of urbanization which combined "four separate indicators of occupation and population-concentration into a single index." Perhaps the major difficulties of the study, however, resulted from the examination of party victories rather than the proportions of the vote and from the concentration on local rather than statewide elections. Significantly, in the one set of state contests considered—the gubernatorial elections—there was "some tendency toward more competition within the more urban counties." Campaigns for major offices that include broad state issues on which urban and rural interests differ may have stimulated more competition than local contests disputed on the basis of less cosmic issues and personalities. Other research (Cutright, 1963) which has challenged the conclusions of the former study in a relatively definitive manner discovered a 14 percent increase in party competition in highly urban compared with relatively rural counties in Iowa.

Perhaps more important than partisan competition in specific urban or rural areas is the potential impact of urban-rural cleavages on conflict within the entire state political system. By providing a basis for enduring political differences and by encouraging parties to battle for votes in a state that contains distinct political interests, urbanization could reshape the political landscape of Iowa. While this development may be retarded by the presence in Iowa of many dispersed medium-sized cities rather than the major urban complexes that exist in many states, the eventual results of urban-rural conflict probably will be determined more by several institutional and cultural characteristics than by purely geographical factors.

Political competition in Iowa probably will be influenced most significantly by the growing Democratic vote in urban centers and by the strength of urban party organizations. While there has been evidence of substantial Democratic gains in the major cities of the state, the continuation of this trend probably will be determined not only by Democratic fortunes at the polls but also by the ability of Republicans to expand their bases of power to include both urban and rural areas. The recognition of this necessity by the Republicans clearly was demonstrated, for example, when party leaders selected a prominent urban lawyer, who had opposed the Shaff plan, as Republican State Chairman one week after the referendum in which the plan was defeated. Prior Republican success in adapting to new political interests and in accommodating fresh movements suggests that renewed Republican efforts to attract the urban vote may have a substantial effect upon party competition in Iowa. In 1968, this Republican chairman, Robert Ray, was elected governor of the state.

If urban-rural divisions are to provide a new basis for party competition in Iowa, they must overcome the legacy that has made Iowa a predominantly one-party state for nearly one hundred years. Traditions usually cannot be repealed, but they can be altered or modified. A powerful heritage may lose some of its resiliency through the development of deep cleavages within the state. Although the development of competitive tendencies in Iowa may depend upon both the astuteness of party leadership and the course of national political trends, they also will be influenced by the dissension created by urban-rural issues and by the electoral strength of urban and rural constituencies.

The voting patterns of various segments of the Iowa population have indicated major shifts in the bases of party strength. An examination of election returns from 1948 through 1964 revealed that small towns have remained steadfastly Republican, while the cities have become increasingly Democratic and farming areas have switched parties erratically in part as their economic fortunes have risen and fallen. The conditions of party competition in Iowa probably will be influenced not only by Democratic gains in the cities but also by the size of the urban vote in state elections. Since population trends have shown relatively steady growth in urban centers and a corresponding decline in rural or small town areas, the changing characteristics of the state could have a decisive impact upon the partisan division of the vote.

The demographic features of Iowa, on the other hand, also might retard the emergence of vigorous party competition by preventing the formation of a unified and cohesive urban force in state politics. Unlike many other states, urban areas in Iowa have not been concentrated in a major metropolis. Instead of one or two large cities that may dominate the economic, social, and political life of a state, urban centers in Iowa are distributed in a number of medium-sized communities located throughout the state. Since those cities commonly have been more Republican than large metropolitan areas, they may fail to develop the solid Democratic majorities that might be necessary for strong party competition in Iowa.

The development of Iowa politics indicates that both the total size of the urban population and the existence of a major metropolis could have an important effect on state political conflict. The accuracy of this proposition, however, cannot be probed without comparative information from all fifty states. This survey is attempted in the next chapter.

While speculation regarding the future course of Iowa politics must be highly tentative, the state seems to have developed the basis for more intensive two-party competition than it has experienced during most of its history. Urban-rural conflict probably has divided the electorate more deeply and more evenly than any earlier dispute in Iowa politics. If Iowa eventually is to be transformed from a one-party to a two-party state, the explanation for this trend might be sought in the pages of United States Supreme Court decisions concerning legislative apportionment. By requiring a careful judicial reappraisal of rural dominance and urban underrepresentation, the courts kindled the aspirations of urban voters and focused attention upon the locus of political influence in Iowa: the state legislature. The court decisions may have had little immediate effect in increasing the probability of urban success in public policy conflicts; but they probably provided an expanded source for the recruitment of state candidates, and they may have aided the Democrats as the party which had been most successful in electing local candidates in urban areas.

In general, emerging trends in Iowa politics probably will be shaped both by the capacity of the Republican party to accommodate urban interests and by the impact of reapportionment and other urban-rural issues upon the political institutions of the state. In the past, Iowa Republicans have demonstrated an impressive ability to incorporate nearly every new political movement within their party ranks. Whether or not Republicans will be able to continue this strategy in the face of Democratic gains in the cities is an issue that must, of course, be resolved from the advantages of historical perspective.

REFERENCES

COLE, CYRENUS. (1921) A History of the People of Iowa. Cedar Rapids: Torch Press.

CUTRIGHT, PHILLIPS. (1963) "Urbanization and competitive party politics." Journal of Politics 25 (August): 552-564.

GOLD, DAVID and JOHN R. SCHMIDHAUSER. (1960) "Urbanization and party competition: the case of Iowa." Midwest Journal of Political Science 4 (February): 62-75.

HAHN, HARLAN. (1963) "Reapportionment, the people, and the courts." Iowa Business Digest 34 (August): 19-22.

___(1964) "Urban versus rural split shows in vote." National Civic Review 53 (March): 146-147.

___(1967) "Turnover in Iowa state party conventions: an exploratory study." Midwest Journal of Political Science 11 (February): 98-105.

HARDER, MARVIN and THOMAS UNGS. (1963) "Notes toward a functional analysis of local party organizations." Unpublished paper presented at the annual meeting of the Midwest Political Science Association, Chicago, Illinois, May 2-4.

LORD, RUSSELL. (1947) The Wallaces of Iowa. Boston: Houghton Mifflin Co.

MASHEK, JOHN R. and DONALD BRUCE JOHNSON. (1954) "Iowa." Pp. 196-216 in Paul T. David, Malcolm Moos, and Ralph M. Goldman (eds.) Presidential Nominating Politics in 1952. vol. 4. Baltimore: The Johns Hopkins University Press.

MATHER, GEORGE B. (1960) A Preliminary Report of an Analysis of the Effects of the Use of Voting Machines in Voting on Special Questions in Iowa, 1920-1956. Iowa City: State University of Iowa Institute of Public Affairs.

PORTER, KIRK H. (1945) "The deserted primary in Iowa." American Political Science Review 39 (August): 732-740.

RYSTROM, KENNETH. (1961) "An Iowa case of gerrymander." Des Moines Register (May 1): 2D.

SALISBURY, ROBERT H., and GORDON BLACK. (1963) "Class and party in partisan and non-partisan elections: the case of Des Moines." American Political Science Review 57 (September): 584-592.

SCHMIDHAUSER, JOHN R. (1963) Iowa's Campaign for a Constitutional Convention in 1960. New York: McGraw-Hill Book Co.

SIMON, HERBERT A., and FREDERICK STERN. (1955) "The effects of television upon voting behavior in Iowa in the 1952 presidential election." American Political Science Review 49 (June): 470-477.

SMITH, PAUL A. (1963) "The local party game: its rules and payoffs." Unpublished paper presented at the National Conference for Education in Politics Seminar, Tuxedo Park, New York, August 31-September 3.

TELFORD, GEORGE B. (1957) "Polk County, Iowa: a study in practical politics." Unpublished paper, Drake University, Des Moines, Iowa.

UNGS, THOMAS D. (1957) "The Republican party in Iowa, 1946-1956." Unpublished Ph. D. dissertation, State University of Iowa, Iowa City.

WIGGINS, CHARLES W. (1960) "The 1960 constitutional convention issue in Iowa." Unpublished paper, Washington University, St. Louis, Missouri.

Chapter 

URBANIZATION AND COMPARATIVE STATE POLITICS

Among the factors that have stimulated public controversies in Iowa, urban-rural distinctions generally have occupied a more common and prominent role than any other feature of the political landscape. Perhaps a careful examination of the politics of any state could not have avoided similar observations. While efforts to disentangle urban-rural differences from other social and economic attributes have been confronted by nearly insurmountable difficulties, disputes between urban and rural regions frequently have seemeed to provide a foundation for significant and persistent conflict. The tendency of many political battles to assume the colors of ethnic, religious, or occupational clashes scarcely has concealed the fact that such groups often have competing centers of influence that correspond to the demarcation between urban areas and the rural countryside. Without the support generated from different locations, the necessary resources for opposing sides in political debates might be absent, and the controversy itself would be deprived of its basis. At various times, urban and rural interests have formed kaleidoscopic networks of agreements and alliances as well as sources of conflict. Such configurations, however, should not obscure the importance of urbanization in the evolution of state political systems.

In fact, the concentration and distribution of the population in a state may have a profound impact on the state's political processes. Perhaps the size of population units, and their attendant rivalries and characteristics, can shape political as well as economic or social development. In order to examine this proposition, it is necessary not only to classify the states by settlement patterns but also to characterize and to identify critical factors in state politics. The utility of a concept can be tested both by its ability to contribute to the detailed explanation of a single case and by its capacity to offer an adequate framework for a broad understanding of comparable phenomena. Presumably, if the urban-rural dimension can provide a satisfactory theoretical perspective for the interpretation of Iowa politics, it should illuminate the analysis of political processes in other states as well. Through such an investigation of American state politics, there is a likelihood that modest but significant progress can be made in formulating a comparative model of general applicability.

In large measure, however, the effort to develop a model for state politics based on urban-rural characteristics has suffered from the limitations of prior research. American state politics has not received the attention from political scientists that would be justified by the scope of the field or its importance (Garceau, 1966; Patterson, 1968). While relatively detailed and extensive scholarly attention has been devoted to even the tiniest foreign nation, only fragmentary and scattered studies have been prepared on many of the state of this country that occupy larger land areas and that govern more people. As a result, principal reliance has been placed upon secondary analyses of individual states in an effort to categorize the frequently complex and heterogeneous characteristics of state political systems.

The field of American state politics not only has suffered from the lack of research on the broad contours of state political processes, but it also has been handicapped by the absence of information on specific state political practices. Many facts about the states that have been commonly unknown conceivably could have been of significant assistance in the formulation of a general model. Consequently, difficulties have been encountered not only in describing relevant aspects of state political processes but also in explaining exceptions or departures from the predominant tendencies. The gaps in the literature about the traditions, personalities, and specific practices that have shaped state politics have created major problems in interpreting the deviant characteristics of states. Until state politics has been explored in greater depth and detail, the construction of a scheme for comparative analysis will be fraught with tentativeness and peril.

The particular method employed in developing this classification of states by stages of urbanization was based upon both the total proportion of persons living in urban places and the presence or absence of large metropolises within the state. Initially, the largest cities in the nation, excluding Washington, D.C., were used as a basis for ranking the states. States were classified by the number of cities they contained of more than 250,000 population, which included the 50 largest cities in the country, and of more than 500,000 population, which identified the 20 most populous cities, according to the 1960 census. Since the level of 10,000 population has been used for defining urban areas, the total urban proportion of the population also was determined for each state. On the basis of both characteristics, the states were classified into eight groups.

The Metropolitan states were from 65 to 75 percent urban, and they all contained cities of more than 500,000 population. These states were among the most populous in the nation, and they included New York, Massachusetts, California, Illinois, and Texas. A group of four states, Michigan, Maryland, Ohio, and Pennsylvania, containing cities of more than 500,000, had a population that was between 50 and 60 percent urban, usually divided in a bi-metropolitan pattern. These were termed the Protometropolitan states. Another group of states was identified which had a population between 65 and 75 percent urban but which included no cities of more than 500,000 population. Consisting of New Jersey, Rhode Island, and Connecticut, these were designated as Suburban states. States that contained a city of either 250,000 or 500,000 and an urban population between 45 and 50 percent were categorized as Interurban. These states included Missouri, Florida, Wisconsin, Minnesota, Oklahoma, Washington, Indiana, and Louisiana. A fifth group of states had a large urban population of from 45 to 65 percent, but their population distribution was shaped primarily by ecological features such as limited space, deserts, or mountains. These states, including Arizona, Hawaii, Colorado, Nevada, New Hampshire, and New Mexico, were called Preurban. States that encompassed a city of more than 250,000 but were only 25 to 45 percent urban included Kansas, Nebraska, Oregon, Tennessee, Virginia, Alabama, Georgia, and Kentucky. These were labeled Midurban states. A population that was between 25 and 45 percent urban but that had no major cities characterized the Transurban states: Iowa, Wyoming, Maine, Montana, Idaho, and Utah. Finally, the states that were less than 25 percent urban and contained no cities of more than 250,000 population were regarded as

Rural. They were North Carolina, South Carolina, North Dakota, South Dakota, West Virginia, Alaska, Arkansas, Delaware, Mississippi, and Vermont.

Unfortunately, the classification scheme represents a relatively static ordering of states at one point in time, since it could not take account of differential rates of urbanization. Clearly, fluctuations in the speed of urban development can have a major impact on the political characteristics that evolve in the states. Such differences, if they could have been incorporated in the taxonomy, might have provided significant explanatory value for interpreting individual deviations from the modal attributes of each group of states. As the classifications demonstrate, however, there is no single line of demarcation between urban and rural or between urban and metropolitan states. Population distributions reveal a variety of patterns and configurations. Furthermore, each type of state partakes, to some extent, in the features of every other type; and the categories perhaps can be regarded as variations along a common dimension rather than as distinct or separate entities.

Another historical factor that may have limited the accuracy of this typology is the influence of sectionalism. At various times, groups of states have been united or divided by distinct regional interests. Although the progress of urbanization perhaps has been largely responsible for reducing the effects of sectionalism in American politics, regional traditions still may occasionally overcome the political similarities spawned by common population attributes. The states included in most of the categories represent widely separated areas of the country, but some differences may be expected due to the intervening effects of regionalism.

Clearly, the principal watershed in the history of American politics was the Civil War. Just as it had a major impact on the development of political patterns in Iowa, the Civil

War also tended to produce distinct political styles and traditions among nearly all the states. As a result, the eleven southern states of the Confederacy may have reflected more divergent political characteristics than most areas, although other regional loyalties also have produced important political differences.

Perhaps more striking than the discrepancies in political characteristics, however, have been the similarities between states at comparable stages of urbanization. For comparative purposes, the following major features of the political processes in each state have been identified: the principal interests that have shaped the history of state politics and the degree of dominance or control that they exercised, the partisan predilections of urban and rural segments of the electorate, the most influential interest groups in contemporary politics, the relative strength or weakness of the governor and the legislature, and the state's experience in reapportionment controversies.

As will be recalled, the history of Iowa politics has been marked by intense conflict between business and agricultural interests that reflected opposing aims of urban and rural districts. In general, Iowa has had a moderately one-party system that has favored a loosely organized but absorbent Republican party. Since the depression, urban areas have reflected moderate Democratic gains, while the small towns have remained steadfastly Republican, and the farm vote has fluctuated widely between the parties. In recent years, lobbying activities in the state have been dominated by business and farm groups, exemplified by the Iowa Manufacturers Association and the Farm Bureau, working in concert rather than in opposition. Iowa also has developed a strong legislature-weak governor system that has tended to perpetuate rural influence. The initiative in the reapportionment controversy stimulated by the Supreme Court decision in *Baker v. Carr* was taken by rural interest groups that sought to maintain their advantage through public and legislative campaigns rather than by urban litigants who attempted to redress their under-representation in the courts. In general, the most prominent features of Iowa politics have paralleled the characteristics of other so-called Transurban states that have begun to emerge from a predominantly rural to a predominantly urban culture.

THE RURAL STATES

The political process in the Rural states differed from this pattern only in a few important respects. Unlike the states that have been undergoing the transitional phase of urbanization, larger communities in Rural states have failed to exhibit the distinct partisan tendencies that normally have been associated with increasing size. Urban areas in Rural states still are relatively small, and perhaps they have not reached the threshold at which a heavy Democratic vote might be expected in the North; but another powerful factor in maintaining party loyalty has been the pervasive influence of small town partisanship in a predominantly rural environment. Until population concentrations have grown substantially, they apparently have had few effects upon a prevailing rural and small town political hegemony. Since the urban base of intense partisan competition has been absent, factionalism within the dominant party often has been a prevalent feature of politics in Rural states. The relatively low stakes of the political game and the homogeneity of the population have produced circumstances in which political contests frequently have revolved about personal factors rather than highly prized economic goals. The propensity of intraparty factional disputes to erupt in the legislature has created a slightly weaker legislative system and a correspondingly somewhat stronger governor than the institutions found in Transurban states. In most other respects, however, the political characteristics of Rural states have been substantially similar to those in Transurban states.

In the history of nearly all Rural states, business interests have occupied a dominant position with occasional strong challenges from farm movements. In Vermont, executives of the marble, railroad, and machine-tool industries have been elected to prominent state offices, although farm groups have formed the principal support for the liberal wing of the Republican party (Lockard, 1959:15,20). Perhaps the most dramatic expression of business-farm conflict originated in North Dakota where the efforts of the Non-Partisan League to wrest political control from the railroads also had a profound impact on the politics of neighboring states (Morlan, 1955; Huntington, 1950). "The most significant aspect of South Dakota's development" has been identified (Clem, 1967:6) as "its involvement in . . . agrarian radicalism." Even in the South, intense competition between business and agricultural interests has been a major feature of the political history of Rural states. In Arkansas, "The great upthrust of organized business killed off the loud but feeble agrarian protest" (Key, 1949:185). Similarly, business interests have enjoyed a major political advantage in South Carolina where alleged champions of the mill workers have failed to support government policies designed to benefit them (Key, 1949:142-145); and "an economic oligarchy has held sway" for half a century in North Carolina (Key, 1949:211). In Mississippi, political conflict has emerged from the divergent economic interests of "the planters of the delta and the rednecks of the hills" (Key, 1949:230). Traditionally, in most Rural states business interests have enjoyed a slight advantage against the sporadic assaults of agricultural movements.

Since the principal battles in Rural states have been fought between business and farm interests within the dominant party, urban areas have had little reason to develop a distinct partisan orientation. In fact, in South Dakota, the largest cities in the state have displayed more support for the Republican party than have rural precincts (Clem, 1967:45, 78-82). Similarly, in West Virginia, the growth of the chemical industry in some of the largest communities of the state has produced an increasing Republican vote (Fenton, 1957:116-117). Due to the Civil War, of course, the dominant party in the South has been Democratic rather than Republican. Yet, there has been little evidence that urban areas in Rural southern states have been more inclined to depart from party orthodoxy than their counterparts in the North. In both areas, the voting behavior of urban centers has stimulated few upheavals in the partisan homogeneity of Rural states.

Perhaps the relatively undisturbed political continuity of Rural states has enabled the same interests that traditionally have dominated political development to occupy a major position in contemporary lobbying. Representatives of agriculture and business have been prominent spokesmen in most Rural states; yet their activities generally have reflected cooperation rather than conflict, and their influence has been relatively strong. In four of the Rural states, interest groups were evaluated by local political scientists (Zeller, 1954:190-191) as "strong," and they were termed "moderately strong" in four other states. None of the Rural states contained a composite of organized interests that were regarded as "weak." In Vermont, two of the principal groups have been the Farm Bureau and the Associated Industries of Vermont, which usually have sought to avoid positions of direct opposition to each other (Garceau and Silverman, 1954:680). Although most legislators have not recognized the lobbyists, "unseen operators" from the two groups frequently have had a decisive impact on state policy (Lockard, 1959:42-43). In Arkansas, the Farm Bureau has been "regarded by observers as a powerful lobbying group," although it often has been joined in common endeavors by the Arkansas Free Enterprise Association (Key, 1949:192). The business and agricultural interests of the delta in Mississippi have been protected by the Delta Council, which has been called (Key, 1949:235-236) "something of a combination of a farm bureau and a chamber of commerce." In many Rural states the joint political objectives of business and agriculture have produced a powerful alliance that has been promoted by increases in the size and corporate nature of farming.

Perhaps the only major exception to the interest group coalitions prevailing in Rural states was found (Fenton, 1957:86) in West Virginia, where the United Mine Workers have established a close affiliation with the state administration. West Virginia also has developed the strongest executive branch of any of the Rural states. Perhaps the special political characteristics of West Virginia have been produced not only by the existence of unique natural resources and the growth of unionism among coal miners but also by the peculiar character of "the Border State environment" (Fenton, 1957:82-125) which required the development of a political organization "to bridge the gap between the Democratic party's conservative or Bourbon and liberal factions." Since a strong central authority was needed to overcome the divisive vestiges of the Civil War, political influence gravitated away from the legislature and the dominant interest has been a mass membership organization that has participated actively in state elections rather than a somewhat more restricted group that focused its political energies on the legislature.

In most Rural states, however, the executive has enjoyed a somewhat stronger position than in the

Transurban states. The relative importance of the governor probably has resulted from the tendency of personal or factional disputes to weaken the influence of the legislature. Party cohesion on legislative roll calls has been absent in many states. In addition, because urban areas have not developed distinct political interests or numerical strength, urban-rural conflict has not been a persistent feature of legislative politics. In Vermont, for example, urban-rural tension in the legislature (Lockard, 1959:39) has been "virtually unknown." Interestingly enough, although rural legislators usually have commanded a majority in South Dakota, they have been on the winning side of "divisive, significant roll call votes" less frequently than the business and professional representatives of the cities or towns (Clem, 1967:110). Apparently, the interests of rural areas never have been sufficiently threatened to necessitate the formation of a unified rural bloc in opposition to urban legislators.

As a result of the diffusion of influence within the legislature, rural governors sometimes have been able to play a relatively strong leadership role. In Vermont, for example, "the governor as the most visible public official and the one with the greatest prestige has a distinct advantage" (Lockard, 1959:43). In South Dakota, one observer (Clem, 1967:86-97) ranked eight of the state's governors as "strong." Although governors of Rural states have not developed the aggressive qualities that have characterized executives in the predominantly urban or metropolitan states, they have exercised somewhat more leadership than might be found in many states with a strong legislative system.

One factor that has undercut the influence of executives in Rural states has been the tendency to recruit governors from the legislative branch of state government (Schlesinger, 1966). In at least five of the Rural states, over 50 percent of the major candidates for state-wide office acquired their first political experience in the legislature. Law enforcement careers provided an entrance into politics for a larger portion of candidates in only two states. Since the legislature has been an important training ground for rural state executives, governors probably have been more reluctant to exert influence upon their former colleagues than office-holders in states that encourage political promotion through other channels. As a result, the predominantly rural interests of the legislature seldom have been disturbed by state-wide elective officials.

Since urban areas have not yet attracted a large proportion of the population, reapportionment has not been a particularly controversial or divisive issue in many Rural states. In fact, legislative resistance to reapportionment has been less prominent in Rural states than in many areas where the size of urban centers has posed a threat to rural domination of the legislature. In some states, such as South Dakota, the legislature has reapportioned its seats relatively frequently in accordance with population shifts (Clem, 1967:101-102). On the other hand, urban growth in North Carolina has provoked several legislative efforts to maintain the rural advantage as well as popular opposition to such attempts (Edsall, 1962:98-109). In general, however, there have been fewer efforts by urban groups to gain equal representation in Rural states than in states where the cities constitute a threat to rural domination of the legislature.

THE TRANSURBAN STATES

Perhaps the clearest test of transitional political development is provided by the states classified as Transurban. Such states not only furnish an opportunity to examine the extent to which their political characteristics parallel those found in Iowa, but they also permit the further identification of trends that may indicate a transition from rural to urban politics. While there have been relatively few prior descriptions of politics in the Transurban states, the general features of their political systems seem to correspond rather closely to the outline of Iowa politics.

Even though the nature of rural organizations has varied widely in the Transurban states, the principal conflict in nearly all of the states has been between business and agricultural associations. In Wyoming, for example, the cattle industry was said to have dominated the life of the state for many years; but the Union Pacific railroad "has been, at least until recently, the most important influence in Wyoming politics" (Peterson, 1940:116-117,120,145-146). While Montana politics has been distinctively shaped by the power of the Anaconda Copper Mining Company and the Montana Power Company, the numerous small farmers who were enticed to settle there by the Northern Pacific railroad have contributed to many of the major battles in state politics (Schlesinger, 1957:87-93; Abbott, 1940). Silver has been "the most important non-agricultural economic interest" in Idaho; but agriculture, timber, railroads, and public utilities also have been "dynamic forces . . . to be reckoned with" (Chamberlain, 1940:173). Similarly, timber companies and hydroelectric utilities, along with textile and shoe manufacturers, have controlled Maine politics for many years (Lockard, 1959:79). Although it has been asserted (Jonas, 1940:17; Jonas, 1959:327-337) that Utah is the preserve of the Mormon Church, religious groups often have followed rather than led "the corporate and financial interests that dominate politics in Utah." Some other states did not experience the same Populist and Progressive battles that erupted in Iowa; but, in nearly all of the Transurban states, the major forces that influenced political development have been business and agricultural interests. Although some type of attack upon business leadership might seem to be an essential background for the emergence of new political trends and movements, a much more critical test of the potential for urbanized politics has been the development of an increasing vote among workers.

Although relatively little attention has been devoted to the analysis of voting behavior in the other Transurban states, there have been some indications of ascending Democratic strength in the cities. In many of the Transurban states, urbanization apparently has yielded a growing labor vote that has produced relatively consistent Democratic margins. The principal strength of the Democratic party in Maine, for example (Lockard, 1959:96), has come from French Canadians and "from the urban workers." The major cities of Utah have established "large segments of Democratic strength" (Jonas, 1957:86). Similarly, in Idaho, Wyoming, and Montana (Martin, 1969:199; Wade, 1969:437; Payne, 1969:204,226-228), urban centers have become sources of Democratic votes. Apparently the emergence of some type of distinct partisan orientation has been a necessary precondition for the transition from a rural to an urban political ethos.

Despite the growth of Democratic sentiment among workers in the cities, interest groups that traditionally have dominated the Transurban states have tended to maintain their influence in the legislature. In addition to the mining companies, agricultural organizations in Montana such as the Farm Bureau, the Farmers Union, the Montana Stockgrowers' Association, and the Montana Woolgrowers' Association (Abbott, 1940:197; Payne, 1969:220-221) have been "the largest group in state political gatherings and, as such, are listened to with respect by political leaders." While the timber lobby in Idaho probably has surpassed "all others in sustained effectiveness," farm groups also have been influential in legislative politics (Chamberlain, 1940:171; Payne, 1969:194). Utilities, the Union Pacific

Railroad, and farm organizations were listed (Huckshorn, 1965:172-174) as lobbies that had the greatest contact with legislators in Idaho. In Utah (Jonas, 1940:35; Jonas, 1969:363) the principal interest groups have been agricultural organizations such as the Farm Bureau and business groups such as "the Utah Manufacturers' Association, the Utah Taxpayers' Association, and the American Mining Congress (Utah chapter)." Although the Farm Bureau and the Grange have been "active in Maine politics," power, timber, and manufacturing clearly have been "the most powerful interest groups in the state" (Lockard, 1959:107-112). In Wyoming, the principal lobbyists have worked for the Wyoming Farm Bureau Federation, the Wyoming Stock Growers Association, and often half of the legislators in the state senate have been members of the Stock Growers Association (Larson, 1965:507-508; Wade, 1969:430-432).

Although most of the Transurban states were listed (Zeller, 1954:190-191) as possessing strong pressure groups, relatively few were judged to have strong party cohesion. In addition, nearly all of the Transurban states have developed political systems containing a strong legislature and a weak executive. The chief executives of Montana, for example, have been "provided with ample constitutional means of influencing the legislature but the tradition of the 'strong' governor has not developed in this state" (Payne, 1954:11). Several Governors of Wyoming (Larson, 1965:511,518) also have stated explicitly that they did not believe the executive had the privilege of interfering with the deliberations of the legislature. In Maine, the strength of the governor has been reduced substantially because the legislature traditionally has appointed the chief administrative officers of the state (Lockard, 1959:105-106). A survey (Huckshorn, 1965:171-172) of legislators in Idaho disclosed ambivalent attitudes toward the governor similar to those found in

Iowa. Although a majority of the representatives believed that the governor had successfully injected himself into legislative decision-making, an equal or slightly larger group felt that acceptance of the governor's recommendations was not necessary for legislative success. Among those who believed that the governor was influential, the largest number cited the threat of the veto as the principal source of his effectiveness, even though he vetoed only 13 out of over 300 bills that were passed during that session. Apparently, evaluations of the legislative role of the governor were shaped both by anticipations of gubernatorial influence and by a desire to maintain the independence of the legislature. Significantly, the legislature has been the major source of gubernatorial candidates in all of the Transurban states, except Montana, where the dominance of an extractive industry apparently has given law enforcement officials an important political advantage (Schlesinger, 1957:87-93). Governors in most of the Transurban states have displayed a striking deference to the legislature. The relative weakness of the governor, therefore, would seem to differentiate rural areas from the states that are progressing toward an urbanized politics.

While there have been few complete descriptions of reapportionment controversies in the Transurban states, the efforts by rural groups to maintain their legislative advantage—that characterized the issue in Iowa—apparently have been made elsewhere. In Wyoming, for example, the Farm Bureau played an active role in an attempt to reapportion the legislature in a manner that would preserve a rural majority (Larson, 1965:536-537). In fact, the activity of groups seeking to continue rural over-representation in the legislature might be an important indicator of states that are experiencing rural-urban transitions. As other bases of political power become increasingly urbanized, rural interests may be expected to make a desperate effort to protect their final stronghold in the legislature.

THE MIDURBAN STATES

In general, the Midurban states have exhibited a mixture of urban and rural characteristics, but their politics have been shaped by a slightly larger urban population and by the presence of a major city. Although the history of Midurban politics has been marked by intense conflict between business and agricultural interests, the Midurban states have developed somewhat stronger partisan differences between urban and rural areas than the Rural states and somewhat weaker distinctions than the Transurban states. The slightly more pronounced partisan cleavages in the latter states probably have resulted from an aroused cognizance of urban interests that was awakened after years of rural domination and that temporarily revitalized party competition. In some Midurban states, however, urban-rural factionalism has developed within the context of a pervasive one-party system. The conditions for business success in lobbying activities have been promoted by the existence of relatively large urban centers which have overshadowed agricultural influence but which have not been sufficient to sustain a strong labor movement. Midurban states also have been characterized by a moderately strong legislature and a relatively weak governor, except in the border states. The importance of urban areas, however, has inspired active litigation to redress the inequities of legislative malapportionment.

Perhaps the sharpest conflict between business and farm interests has occurred in the Midurban states. Kansas and

Nebraska, for example, have experienced nearly parallel political developments both in the nature and intensity of the clash between agrarian movements and the railroads. Virginia politics traditionally has been managed by an oligarchy that has enjoyed "the enthusiastic and almost undivided support of the business community"; but in Georgia "the extraordinary zeal with which the state turned to commercial and industrial development after Reconstruction" produced a bitter and lasting antagonism between agrarian groups and the business interests of the cities (Key, 1949:26-27,118). In most of the Midurban southern states, the controversy engendered by the Populist movement perhaps was even more severe than in the agricultural states of the North. Similarly, in the border state of Kentucky, the principal political onslaughts were directed at the Louisiana and Nashville Railroad and a subsequent "Combine" of business groups by former Populist areas with some labor backing (Fenton, 1957:43).

In most Midurban states, the cities have displayed few of the partisan divergences that might be expected from urban centers of their size, although urban areas in several states have supported important factions within the dominant party. Omaha, for example, has not revealed the Democratic tendencies of other large cities in northern states (Reichley, 1964:74-83). Similarly, the political trend in the urban areas of Kentucky (Fenton, 1957:80) has been "in a plus Republican direction." In large measure, the Republican voting patterns of cities of the Midurban states of the North probably have been molded by a non-industrial environment and by tradition. Cities in the southern states of the Midurban category, on the other hand, have sustained distinctive factions within the dominant Democratic party. In Georgia, the factional alignments produced by Talmagism were built upon a sharp "urban-rural cleavage"; in Tennessee, the political organization erected by

Crump became an urban-based faction (Key 1949:115,58-81). In Virginia, virtually the only opposition to the powerful Byrd oligarchy developed "in the urban industrialized areas" and later in the Washington suburbs; and, in the diffuse atmosphere of Alabama, the large cities have resisted the "friends and neighbors" impulse to support home town candidates (Key, 1949:27,34,40-41). City voters in the Midurban states seem to have recognized the distinctiveness of their political interests, but they apparently have not acquired the electoral strength to challenge the advantage of the dominant party in the state.

The weak and factionalized one-party systems of most Midurban states have created circumstances in which lobbyists can exercise a decisive impact on state politics. The commercial and white collar character of the labor force in the cities also has created a natural base for the power of business groups. In Alabama, for example, business interests have joined with big farmers to quash agrarian radicalism. In Virginia, business and professional men in each community have formed the backbone of the Byrd organization (Key, 1949:33,34,55-56). While business probably no longer exercises the power and influence that lumber companies once wielded in Oregon, it clearly has surpassed the effectiveness of labor or agriculture (Neuberger, 1954:101;Swarthout and Gervais, 1969:313-314). The principal cities of the Midurban states, therefore, have enabled business groups to eclipse farm movements, but they have been insufficient to support a strong labor organization.

While an early study (Zeller, 1954:190-191) of state legislatures indicated that six of the Midurban states has a system of strong interest groups, all except Kansas displayed weak party cohesion in legislative voting. This lack of party loyalty not only has moderated the influence of partisanship, but is also has reduced the authority of the executive. In Nebraska, for example, six former Governors

(Reichley, 1964:82) concluded that the absence of party discipline prevented the governor from exercising strong leadership in behalf of his legislative program. Only in Kentucky and Tennessee, where the balance of forces created by the conflict between the North and the South necessitated a strong centralized administration, have the governors played an effective role in the legislature (Fenton, 1957:14-57; Jewell and Cunningham, 1968:225-226). Significantly, these were also the only two Midurban states in which the governors have been recruited primarily from law enforcement rather than from legislative careers. In most Midurban states, the uneasy equilibrium between rural districts represented in the legislature and the state-wide constituencies of executive officeholders apparently has prevented either branch of government from achieving preeminent influence.

The political importance of the cities in Midurban states, however, has stimulated numerous efforts to end urban under-representation in the legislature. Active attempts by urban residents to secure equal apportionment have been made in nearly all of the Midurban states. In the two states that have strong executives, the governor has assumed a major role in apportionment. In Kentucky, governors frequently have taken the initiative in requesting reapportionment (Jewell, 1962:111-118); while urban officeholders have taken the lead in Tennessee, usually with at least the verbal support of the governor (Crane, 1962:314-325). The size of the urban population may have an important impact on the nature of reapportionment controversies. In the Transurban states, reapportionment frequently has been left to the leadership of rural groups that seek to preserve their advantage. In the Rural states, it has been left to the legislatures which have little to fear from urban representation. The relatively narrow urban-rural division in Midurban states, on the other hand, has inspired urban interests to recognize the potential value of equal apportionment.

THE PREURBAN STATES

Another group of states, the Preurban states, situated somewhere between the urban and metropolitan categories, has a high proportion of urban residents but few major cities. The distribution of the population in Preurban states has been shaped primarily by ecological features such as deserts, mountains, and limited territory. As a result, their politics have partaken of the characteristics of both rural and heavily urbanized states. The Preurban states, therefore, have exhibited some of the heterogeneous attributes not necessarily of a residual category but of a classification that does not occupy a clear position along an urban-rural continuum.

What might be regarded as the normal processes of population growth and settlement in the Preurban states have been inhibited or redirected by geographic boundaries or barriers. Thus, the Preurban states have not been distinguished by any unique or striking political features. Business, labor, and farm interests have played a role in the development of state politics; and the voting patterns of major segments of the electorate have been relatively indistinct. Neither powerful interest groups, a strong legislature, nor an aggressive executive have been able to exert a commanding influence in the Preurban states.

Although each of the three major economic interests have contributed to the history of Preurban state politics, business interests usually emerged as the dominating force. In Arizona, for example, railroad and mine workers succeeded in capturing the state constitutional convention; but they were subsequently quashed and even forcibly expelled by management and large ranch owners, who have enjoyed virtually uninterrupted control over state politics since that time (Waltz, 1940:259-260, 269-270; Reichley, 1964:43). In New Hampshire, railroad and timber interests controlled partisan politics for many years (Lockard, 1959:48); and, in Colorado, organized mining interests traditionally have been "very potent in the state's politics" (Brown, 1940:58; Martin, 1954:62). Nevada, for much of its history, was "used as an agency of the Southern Pacific Railroad, and as the plaything of San Francisco nabobs" (Ostrander, 1966:132). In some Preurban states, the political forces of farming and business were combined to such an extent that they were virtually indistinguishable. The sugar industry in Hawaii, for example, acted for many years as a commercial interest rather than as an agricultural operation (Fuchs, 1961). Similarly, in New Mexico, farmers, ranchers, and business combined to form "the dominant economic interests in the politics of the state" (Donnelly, 1940:229; Stumpf and Wolf, 1969:280). While the natural resources of the Preurban states have permitted business, agriculture, and labor to participate in political development, the small urban concentrations in the states have seemed to provide a better basis for business influence than for agricultural or labor strength.

Urban centers in the Preurban states have not developed the distinct partisan preferences that might be anticipated in states with a high proportion of city residents. The relatively small cities in New Hampshire, for example, have been only slightly more Democratic than rural areas; but Democratic gains have been closely associated with increasing industrialization (Lockard, 1959:63-64). Arizona Republicans, on the other hand, seem to have been gaining more rapidly in Phoenix than in other parts of the state

(Reichley, 1964:52; Bushnell, 1969:253-254). The urban vote in Nevada has been divided between Reno which is normally Republican and Las Vegas which is normally Democratic (Smith and Kunsman, 1954:102). Even the major city in the Preurban states, Denver, has been "predominantly but not dependably Democratic" (Martin, 1954:63; Irwin, 1962:73). Urban areas in the Preurban states, therefore, have not seemed to develop the distinct interests that would differentiate their voting patterns from other parts of the state (Holmes, 1967:266-267).

Although business interests probably have established the most effective lobbying organizations in most of the Preurban states, the strength of the groups has been variously defined (Zeller, 1954:190-191). In Arizona and New Mexico they were regarded as "strong," in Nevada they were labeled "moderate," and in Colorado they were considered "weak." General agreement (Lockard, 1959:70) seemed to exist in New Hampshire that the three most powerful interest groups were "the race tracks, the public utilities, and the lumber-manufacturing interests." In Nevada, land, livestock, mining, and business interests, as well as the Farm Bureau, have been listed (Wier, 1940:107-108; Smith and Kunsman, 1954:102) as the most influential groups in the state; while in New Mexico this list (Donnelly, 1940:242; Holmes, 1967:193-194) has included the Taxpayer's Association, the Cattle Grower's Association, and the chambers of commerce. In Arizona, the power utilities apparently have replaced the copper mines "as the most active lobbyists in the state capitol" (Reichley, 1964:48; Rice, 1969:57). With a few exceptions, therefore, the most influential lobbying groups have not represented the largest economic segments of the population. Since the highly regarded interest groups in Preurban states generally have been confined to limited objectives and have not necessarily included the groups that could mobilize the greatest popular support, the effect of lobbyists has been labeled "moderate" rather than "strong."

While neither the legislature nor the executive has acted aggressively to provide political direction in the Preurban states, there have been growing indications that the governor may fill this role. In New Hampshire (Lockard, 1959:73), for example, his authority has not been "comparable to that of some governors . . . but the trend in New Hampshire is toward greater gubernatorial power." Similarly, in Arizona, one observer (Morey, 1965:113-114) concluded, "At present the governor's role as legislative leader is comparatively weak, yet there are factors at work which will no doubt provide him with more strength." If the governors of Preurban states are to achieve a major position of legislative influence, they probably will have to increase the party cohesion of legislators, which was judged strong only in Colorado (see, however, Gomez, 1969:142-143). Yet the strength of governors may be enhanced by their political independence from the legislature. In all of the Preurban states, except New Hampshire and New Mexico, more governors have been recruited from law enforcement than from legislative careers. As the population of sparsely settled areas declines and urban centers grow, state-wide elective officials rather than legislative representatives may emerge as the most influential force in state politics.

Despite the political importance of urban centers, controversies regarding reapportionment have not been as severe in the Preurban states as they have been elsewhere. Malapportionment in New Hampshire, for example, has not been extreme (Lockard, 1959:73). Similarly, in Colorado (Irwin, 1962:68), "the public, even the urban public, had expressed no great indignation on the subject." The relative equality of apportionment in the Preurban states apparently has been produced not only by the high proportion of urban residents but also by the moderate importance of the legislature which has not been an exclusive preserve of rural interests. The politics of Preurban states, therefore, has reflected an unusual pattern of mixed urban and rural influence shaped by geographic rather than population attributes.

THE INTERURBAN STATES

A clearer example of the impact of urban-rural characteristics on state politics has been provided by the Interurban states. The relatively even division between urban and rural populations as well as the presence of major cities has created a setting in these states in which business, farm, and labor interests have competed on an equal and intense basis for the rewards of political influence. This struggle frequently has been evident in state elections where voting patterns have reflected sharp partisan differences between urban and rural areas. Although many of the battles have been waged by groups outside the government, the strength of the party organizations has offered the governors a major opportunity to exert political leverage. Urbanization, therefore, has provided a firm basis for vigorously competitive politics in the Interurban states.

Perhaps no other group of states has contained a richer variety of agricultural, business, and labor interests in contests for political supremacy. In Minnesota, for example, the combined attacks of agrarians and workers upon the power of the railroads were successfully united in the Farmer-Labor party, which continued to provide a basis for a renascent Democratic party in the post-war era (Mitau, 1960:3-28). Similarly, in Oklahoma, urban and rural opposition to oil and other business interests was joined in the Farmer-Labor Reconstruction League (Debo, 1949:47).

Support for the Progressive movement fashioned by Robert M. La Follette in Wisconsin was provided by both Scandinavian farmers and by workers in Milwaukee (Epstein, 1958; Weibull, 1965). In Washington, the survivors of the populist movement "continued to work through the Washington State Grange and labor organizations to counteract the railroads and other big corporations" (Avery, 1965:217). In Louisiana, of course, Huey Long led a tattered band of the underprivileged against Standard Oil and other financial interests (Key, 1949:156-182; Sindler, 1956). Only in Florida and Indiana were business concerns able to gain a relatively decisive victory (Havard and Beth, 1962:26; Fenton, 1966:162). Although both states had a colorful background of agrarian discontent, the emotions engendered by the Civil War seemed to prevent a showdown between the major economic interests. In most Interurban states, however, the characteristics of the population not only provided a basis for significant agricultural and labor movements but they also allowed the two groups to form a coalition against the ruling business interests.

In large measure, the same groups that historically have battled for political power in the Interurban states have continued to exercise a major influence in contemporary politics. In Washington, for example, the Grange has cooperated with organized labor in opposition to business interests which have been most aggressively promoted by the electric power companies (Baker, 1960:4; Bone, 1969:400-401). Although the most influential lobbyists have drawn support primarily from rural legislators, a list (Havard and Beth, 1962:222-225) of the 17 most powerful lobbies in Florida included 15 business interests and only two agricultural groups–but no labor unions. The most successful interest groups in Minnesota have included the Minnesota Employers' Association and the Farm Bureau, which have enjoyed the support of the Republicans or

Conservatives, while the Farmers Union and the AFL-CIO have relied upon Democrats or Liberals (Mitau, 1960:85-97). A similar rating of effective interests and a similar split in agricultural groups apparently has developed in Wisconsin (Epstein, 1958:20-21). One of the few groups that has preserved both its influence and the relative unity of agrarian interests has been the Missouri Farmers Association (Fenton, 1957:143-145). The high stakes of politics in the Interurban states, therefore, have promoted a continuity and a balance among major economic interests that usually has prevented any one group from exercising complete dominance.

The even distribution between urban and rural segments of the population in Interurban states not only has stimulated intense competition among major economic interests, but it also has promoted clear partisan disagreements between urban and rural areas. Perhaps the most extensive research on voting behavior has been conducted in Wisconsin (Epstein, 1956; Epstein, 1958:57-76; Adamany, 1964). Not only have the voting patterns of cities, small towns, and farms in the state closely paralleled those found in Iowa, but they also revealed a substantial increase in the Democratic vote in the major city, Milwaukee. Similarly, in Minnesota, "the increases in the cities were a significant contributing factor to Democratic-Farmer-Labor electoral success" (Fenton, 1966:96; Mitau, 1960:29-33). In Missouri, St. Louis has joined Kansas City as a major Democratic stronghold (Fenton, 1957:141, 153-154). Growing Democratic margins in the cities of Indiana also have undercut the old sectional divisions that were established by settlement patterns during the Civil War era (Fenton, 1966:179-180, 184-189). Even in the one-party environment of the South, factional alignments have been shaped by urban-rural distinctions. In Florida, the traditional cleavages between northern and southern portions of the state have not reflected "ordinary sectionalism: they are in fact urban-rural differences" (Key, 1949:92). Opposition to the Long organization in Louisiana also has been associated with urbanism (Sindler, 1956; Key, 1949:176-177). The political interests of urban and rural segments of the population apparently have been recognized by the voters in the Interurban states and expressed in sharply divergent partisan preferences.

Although the political influence exerted by the parties and interest groups probably has surpassed the power wielded by government officials, there have been growing indications of increasing gubernatorial authority in the Interurban states. Except for Minnesota, Oklahoma, and Florida, the governors have been recruited from law enforcement rather than from legislative careers. In the vacuum created by the nonpartisan legislature in Minnesota, however, "the governor preempted representation of the interests of the people" (Fenton, 1966:108). Governors of Wisconsin also have acquired considerable prestige by virtue of their ability to satisfy the demands generated by protest movements (Epstein, 1958; Fenton, 1966:68-69). In two Interurban states, however, governors have been handicapped by constitutional limitations. Missouri governors, for example, have been deprived of administrative responsibility by numerous boards and commissions and of legislative leadership by the prohibition against a second term (Fenton, 1957:135-136); and, in Florida, gubernatorial powers have been curbed by a "cabinet" government which has granted various administrative duties to six elected state officials (Key, 1949:99). In most of the other Interurban states, however, the governor has captured the opportunity for political leadership as the principal representative of a heterogeneous population.

The narrow division of the population between rural and urban areas as well as the competition between major economic interests in the Interurban states probably has produced a political system that approximates the democratic model more closely than in many other states. Significant segments of the population have recognized the importance of politics and have prevented any one interest from dominating the contests. Although the legislature frequently has been an interested spectator rather than the crucial participant in the struggle, reapportionment has been a source of active controversy and urban initiated litigation in most Interurban states. The close urban-rural balance as well as the existence of a major city in the Interurban states, therefore, has provided the basis for a kind of aggressive politics that many observers have desired in rural or metropolitan states.

THE SUBURBAN STATES

Just as an encircling network of suburbs has become a common feature of metropolitan areas, there is a small group of states in the megalopolis of the eastern seaboard that has exhibited the characteristics of suburban politics. These states all have large urban populations, but they contain no major cities of 250,000 or 500,000 population. In many respects, they have seemed to mark a buffer zone between urban and metropolitan states.

As might be expected in states that have not contained a major city or a large rural population, business interests traditionally dominated political development. In Rhode Island, business was protected by the small town Republican machine headed by General Charles Brayton, who served as a counsel for the railroads, utilities, and "other large corporations wanting favors from the state." A similar machine was led in Connecticut by J. Henry Roraback, who acted as a railroad lobbyist and eventually became the president of the Connecticut Light and Power Company (Lockard, 1959:176,245-251). In New Jersey business interests were dominated by the railroads. One observer (Lockard, 1964:57-58) noted that "railroads from Maine to California played an important role in state politics during the nineteenth century, but in no state was that role assumed earlier or more pervasively than in New Jersey."

Gradually, however, there have been indications that this hegemony has begun to collapse. Democratic gains in the larger cities have yielded a revitalized competition between the political parties. Connecticut, for example, has become "divided like Caesar's Gaul into three parts—the larger cities, the rural small towns, and the new suburban retreats" (Lockard, 1959:234). Although the Republicans have maintained control of the small towns and suburbs, urbanization has stimulated both enhanced partisan competiton and the representation of new political interests.

Ultimately, both labor and management have emerged to battle for the position of the strongest interest group in the Suburban states. An early study of pressure groups in New Jersey (McKean, 1967:104-115), for example, concluded that the State Federation of Labor and the Manufacturers' Association which opposed the unions were the strongest interests in the state, although "over a period of time the Federation will win because of its great voting strength." A later investigation (Wahlke et al., 1962:313-323), however, concluded that business still ranked first and labor rated second among the major economic interests in the state. In Rhode Island labor has become "a vitally important element in the legislature," although the race tracks and other business interests also have commanded sizable support. Interest groups in Connecticut have tended to divide along partisan lines (Lockard, 1959:222-225,285-288). While labor has been consistently aligned with the Democrats, the Connecticut Manufacturers Association, farm groups, and the insurance lobby have been linked with the Republicans.

Strong party organizations in Suburban states have given governors an opportunity to exert effective legislative leadership. All three states possessed strong party cohesion in legislative voting. "In Connecticut the party leadership is the real leadership of the Legislature." Similarly, in Rhode Island, the governor and the party leaders usually have controlled the legislature (Lockard, 1959:212,278). Another assessment of New Jersey governors (Lockard, 1964:101) concluded that "the modern governors in varying degrees have had party support because they became party leaders as governor; they were more than titular leaders of their parties–when they were most successful they were the driving and dominating figures in their parties."

As might be expected in states that have experienced relatively sudden growth, the apportionment of the legislature has failed to reflect new elements of the population or new political alignments (see Shank, 1969). In Rhode Island, for example, small town Republicans continued for many years to maintain control of the state senate. Similarly, the over-representation of small towns in the Connecticut House has been one of "the worst of them all" (Lockard, 1959:178-179,272). Since the expanding suburbs frequently sympathized with the political aims of small town Republicans, there was little impetus for reapportionment until the battle could be waged primarily in the courts rather than in the field of electoral politics.

THE PROTOMETROPOLITAN STATES

While suburban politics has been distinguished from the politics of the metropolis, there is another group of state whose political styles clearly have been influenced by a large rural population contained within the states. The Protometropolitan states all include a major metropolis but a total population that is only 50 to 56 percent urban. Their urban centers, therefore, have been too large to facilitate equal competition among major economic interests, but not large enough to dominate the politics of each state.

The relatively late development of labor as an effective political force and the preeminent position of the cities, which overshadowed agricultural strength, left an unobstructed path for business domination in the history of most Protometropolitan states. In Michigan, the pervasive power of the Republican party was sustained by alliances with railroad and other business interests and later by association with the automobile manufacturers (Sarasohn and Sarasohn, 1957). In Pennsylvania, also, the Republican party long was supported by wealthy families, business-led city machines, and the Pennsylvania Manufacturers Association (Reichley, 1964:54-61). Business interests in Ohio were promoted by both the Republican and Democratic parties (Fenton, 1966:122); and, in Maryland, the Democrats were backed by business interests in Baltimore and by large landowners in the Tidewater area (Fenton, 1957:171).

Gradually, however, the prevailing one-party hegemony was disrupted by sharp partisan differences. Since the Great Depression, metropolitan counties in Ohio have become increasingly Democratic, while the rural corn belt counties have reflected growing Republicanism (Fenton, 1966:144-145; Flinn, 1960). Furthermore, in Baltimore, 1932 marked a shift from Republican to Democratic margins (Fenton, 1957:182). In Michigan the Detroit area has switched completely to the Democratic fold, although the medium-sized cities where working class cohesion was reduced have remained at least partially Republican (Fenton, 1966:31-33; Masters and Wright, 1958). The Republican city machines in Pennsylvania, on the other hand, were not completely destroyed until the 1950s (Reichley, 1964:61-66).

The electoral dislocations produced by the growth of working class populations in the Protometropolitan states also has been reflected in the position of interest groups. The political parties in Michigan have been "backed by the two greatest concentrations of economic power in the state, the major automobile manufacturers and the United Automobile Workers of America" (Sarasohn and Sarasohn, 1957:68). In Pennsylvania one observer (Sorauf, 1963:52) concluded, "Popular myth–and that myth is widely affirmed in party circles, too–has it that the Pennsylvania Manufacturers Association and the Pennsylvania Railroad exercise inordinate power in the councils of the Republican party." A survey of legislators in Ohio (Wahlke et al., 1962:314) revealed that 42 percent referred to business interests as the most salient lobby, while labor was the second most frequently mentioned economic interest, having been cited by 12 percent. In all of the states, except Michigan, however, the strength of pressure groups was labeled (Zeller, 1954:190-191) "moderate."

The conflict between economic interests as well as between parties has created circumstances in which the governor can exercise effective leadership through strong party cohesion in the legislature. Even in Ohio, for example, "gubernatorial initiative in legislative policy" has been "accepted as right and proper" (Wahlke et al., 1962:386). Significantly, law enforcement rather than legislative careers have been the principal source of gubernatorial candidates in all of the Protometropolitan states except Maryland. Thus, the opportunities for executive leadership in these states have been as fortuitous as in most of the urban or metropolitan states.

The political prominence of metropolitan areas in states

that contain large rural territories also has created a drive for reapportionment unsatisfied by litigation in the courts that involved most of the major political interests as well. Political parties and interest groups played a major role in the struggle over reapportionment in Michigan (Lamb, 1962); in Maryland, specially organized citizens' groups were a driving force for reapportionment (Pettengill, 1962:298-313). The struggle over reapportionment in metropolitan states, therefore, has been of such a critical nature that it has enlisted the support of the major urban political movements in the state.

THE METROPOLITAN STATES

The final form of political development in America probably has been represented by the Metropolitan states. These states, which contain the decisive electoral votes in presidential elections, also have included the principal cities and the largest population concentrations in the country. Since the Metropolitan states may reflect the ultimate nature of a political system toward which the other states have been progressing, considerable attention will be devoted to the characteristics of politics in the most populous states.

Although the relatively late emergence of labor as a politically potent organization created a vacuum in which business could exercise dominance, the large cities also enabled labor to acquire political strength earlier and more extensively in the Metropolitan states than elsewhere. Perhaps the strongest alliance between labor and the Populist movement, for example, was formed in Illinois to oppose the railroad, stockyard, and utilities interests that "ran the state almost as they pleased" (Nye, 1965:88). Similarly, in Texas, which also experienced the impact of an early Populist-business conflict, the principal basis of the liberal-conservative dichotomy has been economic (Soukup et al., 1964:14). Oil and other major business interests have been opposed by a loose combination of the Texas AFL-CIO, Negro and Latin American groups, and independent political organizations or leaders. In California the Southern Pacific Railroad controlled the state as a virtual political fiefdom until the Progressive movement of the early twentieth century (Mowry, 1963). In New York the Republican party and the state were run by the utiltites and other big business interests until labor acquired political strength during the Great Depression (Moscow, 1948:71). In addition, Massachusetts politics long was controlled by influential families representing the leading business concerns of the state. Although the principal political conflicts in New York (Glazer and Moynihan, 1963) and Massachusetts (Levin and Blackwood, 1962:28-29) often have been portrayed as ethnic rather than business-labor clashes, the struggle between the native Yankees and the later immigrants usually seemed to reflect a basic economic disagreement between workers and management. Massachusetts, in fact, allegedly has evolved a technological economy which may presage developments in other states. As a result, the principal political division in Massachusetts has become "the cleavage between the workers and the managerial leadership" (Litt, 1965:17).

The political conflict that has been emerging between labor and management in the Metropolitan states also has been reflected in the ranking of major economic interests. In New York the two lobbies that "were peculiarly effective in Albany over a period of years" were the New York State Federation of Labor and the Associated Industries of New York State (Moscow, 1948:203). Economic interests in Massachusetts have been divided along partisan lines. Labor has been one of the main groups in the Democratic party, and the Republican party has been closely aligned with "the public utilities interests, the real-estate lobby, the Associated Industries of Massachusetts, the Chamber of Commerce, the insurance companies, and the Massachusetts Federation of Taxpayer's Associations" (Lockard, 1959:163-165). Major legislation in Illinois frequently has been the result of compromises negotiated between labor and management lobbyists (Steiner, 1951). Although the oil industry clearly has been the dominant interest in Texas, its major opposition has come from a coalition led in part by labor unions (Soukup et al., 1964). Only in California, where the parties have been curbed by legal inhibitions and interest groups have been exceptionally strong, has major influence been wielded by individuals representing numerous smaller industries rather than by large economic interests (Buchanan, 1963). In most of the states, however, the principal legislative battles have been waged by labor unions and business organizations. As the most economically advanced areas in the nation, the Metropolitan states probably represent the modern industrial state in which the conflict between workers and management has become so intense and so divisive that it has absorbed or inhibited many other types of political disputes.

Despite the prominence of cleavages between businessmen and workers, many of the larger cities have not developed the distinctive partisan preferences that might be expected of urban centers. Perhaps the absence of a heavy Democratic vote in populous towns has resulted from the tendency of one or two major metropolitan areas to dominate the politics of the state. Rural areas and suburbs in New York, for example, have been as steadfastly Republican as the city has been Democratic; but the large upstate towns such as Albany, Buffalo, Rochester, and Syracuse have displayed "a much higher degree of Republicanism than is normally expected of industrial regions in the United States." Similarly, in California, large towns such as

San Diego as well as satellite cities in metropolitan areas have been Republican; and many of the state's Republican legislators have represented heavily urbanized areas. The major cities in Texas also have contributed the largest proportion of the Republican vote in that state (Reichley, 1964:98,118-121,171,188). Since 1932 in Illinois, only the two counties containing Chicago and the Rock Island-Moline area have become consistently Democratic, along with the downstate counties whose Civil War loyalties were reinforced by urbanism or coal mining (Fenton, 1966:195-196). Other populous urban centers have exhibited few Democratic tendencies.

Perhaps the most thorough analysis of voting behavior in any Metropolitan state, however, has been conducted in Massachusetts. Although the towns dominated by Yankees initially were more Republican than the cities settled by immigrants, Boston traditionally has provided the heaviest Democratic margins in the state and the Democratic vote has increased in relation to the size of cities (Huthmacher, 1959:272-278). Foreshadowing what may become a trend in other areas that contain a large technical and professional labor force, however, the largest cities gradually have contributed a declining proportion of the total state vote and less of the Democratic vote than the suburbs (Litt, 1965:47-51). Although the principal metropolis in the Metropolitan states usually has exhibited party preferences quite different from the choices made in the remainder of the state, many of the other large urban areas have not shared the partisan predilections of the metropolis. In part, such circumstances may have been promoted by the rivalry stimulated by the presence of a massive and overpowering metropolis. In any event, the clear and consistent urban-rural differences that might be anticipated in urbanized states have not always been evident at the extreme end of the metropolitan spectrum.

Although legal obstacles reduced party cohesion in the

California legislature for many years, partisan competition in most of the metropolitan states has enabled the governor to exert effective leadership. In one-party Texas there has been widespread factionalism and little urban-rural conflict except on "such issues as reapportionment of legislative seats, reduction of farm-to-market road programs, and, to a lesser degree, utilization of water resources" (Soukup et al., 1964:81). In Illinois, where there has been moderate cohesion and little evidence of urban-rural disagreement, the governor "has more opportunities to influence legislative action than any other outside source" (Steiner and Gove, 1960:33). The system of a "pre-veto" and the threat of a formal veto has given the governor of New York a strong role in the legislative plans of both parties (Moscow, 1948:175-179). Although Massachusetts has been the only Metropolitan state in which gubernatorial candidates have been recruited primarily from legislative rather than law enforcement offices, one commentator (Lockard, 1959:159) has concluded, "Through his formal power as chief executive and his informal power as chief of the party, the governor has to be considered the most important single person in making legislative decisions."

Despite the significance of the governor in the politics of Metropolitan states, efforts to obtain legislative reapportionment have expanded beyond his authority to enlist the support of many other important political interests. In New York, where the initial suit to secure reapportionment was brought by a city radio station, the issue became a major partisan controversy (Lee, 1967). Various referendums on reapportionment in California also generated the activity of party leaders and economic interests (Baker, 1962:51-63). Similarly, in Illinois, a constitutional amendment on apportionment in 1954 attracted the support of farm, business, and labor organizations as well as the political parties (Steiner and Gove, 1960:89-90). A 1961 reapportionment measure in Texas aroused the interest of labor unions, but its outcome was largely determined by factional alliances in the legislature (Cherry, 1962:120-130). Perhaps the only Metropolitan state that has maintained an equitable apportionment of legislative seats is Massachusetts, where the suburbs have displayed more support for the political goals of the cities than elsewhere (Lockard, 1959:152). In most of the Metropolitan states, however, the political stakes of the reapportionment controversy have been sufficiently important to stimulate the commitment of the major urban interests.

The political resources created by urban or rural populations within a state apparently have had a critical impact upon state politics. Although the evidence has suggested that state political systems reflect a continuous distribution rather than completely separable units, both the presence or absence of a major city and the proportion of urban residents seem to have shaped the political characteristics of the states. Within each category, as the states have become increasingly urbanized, they have displayed attributes that are both relatively uniform and that are somewhat distinct from states experiencing an earlier stage of urbanization.

Since the influence of economic interests upon political development has been determined largely by historical factors rather than by contemporary population characteristics, the principal conflicts in the states seemed to have relatively little effect on the evolution from rural to urban to metropolitan politics. Strong conflict between business and agricultural interests was evident in Rural, Transurban, and Midurban states. A somewhat indistinct pattern involving business and labor as well as farm groups began to emerge, however, in the Preurban states. This trend was confirmed by strong competition between the three groups in the evenly divided Interurban states. Nonetheless, the relatively late emergence of labor as a potent political force created nearly ideal conditions for business domination in both Suburban and Protometropolitan states, where agrarian movements were repressed by the political strength of the cities. At the end of the spectrum, however, the Metropolitan states seemed to provide a sufficiently massive urban base that both restricted the impact of farm groups and that sustained the gradual growth of labor influence. As a result, the political history of Metropolitan states has been characterized by moderate conflict between business and labor organizations.

Although the impact of population attributes perhaps has been more apparent on other political variables, the urban or rural nature of the states seemingly has affected the role of economic interests in political development. In addition, the examination of state political histories clearly has demonstrated the critical importance of business interests in nearly all sections of the country.

Perhaps the clearest indications of the rural-urban transition in state politics has been evident in the voting patterns of important segments of the population. In the relatively homogeneous environment of the Rural states, there were few differences in the partisan preferences of cities and rural areas. An important distinction, however, was revealed in the Transurban states where the cities began to develop moderate Democratic tendencies. Although the one-party nature of many Midurban states partially reduced the partisan conflict between cities and rural areas, a similar pattern of only slight divergences in the voting behavior of urban and rural places was found in Preurban states. In the highly competitive Interurban states, strong disagreements were found in the electoral choices of cities, small towns, and farms. The Republican propensities of the suburbs inspired a moderate party cleavage in Suburban states that apparently overlapped into the Protometropolitan states. In the advanced Metropolitan states, however, many large cities have displayed less partisan independence from rural or suburban areas then might have been anticipated. In part, the rivalries engendered by the party dominance of a major metropolis may have inspired this phenomenon. Despite this strange reversal of a well-established pattern of urban and rural voting behavior, however, urbanization seemed to have a pronounced impact upon electoral alignments. Not only have urban and rural areas produced different party majorities, but population attributes also seem to have molded the intensity of voting patterns within the states.

In addition to their effect upon voting behavior, the urban-rural characteristics of states seem to have had a significant impact upon interests groups that have emerged as major contenders for political influence. The same groups that traditionally have shaped the development of politics in many states have retained their strong positions in modern lobbying efforts. In both Rural and Transurban states, for example, the major pressure groups represented business and agricultural interests that frequently have acted in concert rather than in conflict with each other. On the other hand, agricultural or labor organizations have ebbed sufficiently in Midurban or Preurban states to give business interests preeminent influence. In the industrialized states, however, labor unions have challenged business organizations to a contest that has been relatively moderate in Suburban and Metropolitan states and strong in Protometropolitan states.

The demographic attributes and configurations of interests within states also have played a role in shaping executive-legislative relations. Factional tendencies in Rural state legislatures, for example, have seemed to give governors a slight edge in dealings with the legislature. The Transurban and Midurban states have developed a tradition that has strengthened rural legislative interests and has partially undermined the effectiveness of governors who must be elected with a sizable urban vote. In most of the remaining states, however, the governor has enjoyed an advantage in legislative leadership that has ranged from weak in the Preurban states to strong in the Metropolitan states. Despite the strong legislature-weak governor systems in states that have begun to emerge from rural-dominated politics, therefore, the trend in most urban and metropolitan states has been toward increasing executive influence and authority.

A final test of the impact of urbanization on state politics has been provided by the controversy regarding legislative reapportionment. In Rural states, where the size of the cities has not constituted a danger to rural control of the legislature, apportionment usually has been prompted by legislative initiative, and malapportionment has not been as prevalent as elsewhere. Rural interests in the Transurban states, however, have recognized the potential threat of increased urban representation, and they have sought to preserve their legislative strength. In most of the other states, reapportionment has been attempted primarily through urban-sponsored litigation in the courts, except in the Protometropolitan and Metropolitan states where the apportionment issue has been so pressing that it has attracted the attention of major economic, partisan, and other organizations.

A comparative examination, therefore, not only has indicated the feasibility of developing a model for the analysis of state politics but it also appears to have provided some evidence for the broad proposition that urban-rural attributes may have an important impact on state political systems. The concentration or distribution of population within a state may have a more significant effect upon politics than has been commonly realized. A survey of legislators in four states (Wahlke et al., 1962:423-427) found, for example, that urban-rural disagreement was regarded as the principal cleavage in both houses of the Ohio legislature, in the California House, and in the Tennessee Senate. It was considered the second most divisive force in the California Senate; and it was in third place in the Tennessee House and in both chambers of the New Jersey legislature. In most places, it outranked partisan conflict, ideological debates, regional divisions, contests between supporters and opponents of the governor, and economic clashes between labor and it enemies. While urban and rural features may be closely related to other significant attributes, they have often seemed to shape the nature of political controversies and to provide the basis for political conflict.

By examining the broad contours of politics in all states as well as the detailed characteristics of a single state, this study has sought to provide not only a comprehensive description but also an assessment of major directions in state politics. Although the prominent urban and metropolitan states have been most impressive because of their size; they probably have revealed less information about the process of political development than areas that have been in the process of transition from urban to metropolitan or particularly from rural to urban politics. States such as Iowa have not represented "political museum pieces," as Key (1949:19-35) described Virginia, but they have formed important cases that might illuminate the differentiating features of state politics generally. In Iowa observations have been made of the early conflicts between business and agricultural interests, that emerged as other economic clashes occurred in more industrialized states, and of the initial partisan stirrings in the electorate that may eventually revitalize party competition. In addition, the state has yielded an example of the domination of lobbying activities by cooperating business and agricultural interests, a partnership that may collapse as the farm population declines and as labor begins to grow. Perhaps the clearest indications of possible political trends, however, have been provided by the strong legislative system that may be undermined by aggressive governors–supported by urban voters–and by the struggle over reapportionment and the desperate attempt of rural groups to perpetuate their legislative advantage. In general, therefore, the specific features of Iowa politics have seemed to provide an adequate benchmark for the analysis of state political processes. The transition from rural to urban politics has reflected both the background and future trends of American state politics.

REFERENCES

ABBOTT, NEWTON CARL. (1940) "Montana: political enigma of the northern rockies." Pp. 189-217 in Thomas C. Donnelly (ed.) Rocky Mountain Politics. Albuquerque: The University of New Mexico Press.

ADAMANY, DAVID. (1964) "The size-of-place analysis reconsidered." Western Political Quarterly 17 (September): 477-487.

AVERY, MARY W. (1965) Washington: A History of the Evergreen State. Seattle: University of Washington Press.

BAKER, GORDON E. (1960) The Politics of Reapportionment in Washington State. New York: McGraw-Hill Book Co.

——(1962) "The California senate: sectional conflict and vox populi." Pp. 51-63 in Malcolm E. Jewell (ed.) The Politics of Reapportionment. New York: Atherton Press.

BONE, HUGH A. (1969) "Washington state: free style politics." Pp. 381-415 in Frank H. Jonas (ed.) Politics in the American West. Salt Lake City: University of Utah Press.

BROWN, ROY E. (1940) "Colorful Colorado: state of varied industries." Pp. 51-87 in Thomas C. Donnelly (ed.) Rocky Mountain Politics. Albuquerque: The University of New Mexico Press.

BUCHANAN, WILLIAM. (1963) Legislative Partisanship. University of California Publications in Political Science, vol. 13. Berkeley: University of California Press.

BUSHNELL, ELEANORE. (1969) "Nevada: the tourist state." Pp. 233-257 in Frank H. Jonas (ed.) Politics in the American West. Salt Lake City: University of Utah Press.

CHAMBERLAIN, LAWRENCE HENRY. (1940) "Idaho: state of sectional schisms." Pp. 150-188 in Thomas C. Donnelly (ed.) Rocky Mountain Politics. Albuquerque: The University of New Mexico Press.

CHERRY, H. DICKEN. (1962) "Texas: factions in a one-party setting." Pp. 120-130 in Malcolm E. Jewell (ed.) The Politics of Reapportionment. New York: Atherton Press.

CLEM, ALAN L. (1967) Prairie State Politics. Washington: Public Affairs Press.

CRANE, WILDER. (1962) "Tennessee: inertia and the courts." Pp. 314-325 in Malcolm E. Jewell (ed.) The Politics of Reapportionment. New York: Atherton Press.

DEBO, ANGIE. (1949) Oklahoma: Foot-loose and Fancy-free. Norman: University of Oklahoma Press.

DONNELLY, THOMAS C. (1940) "New Mexico: an area of conflicting cultures." Pp. 218-251 in Thomas C. Donnelly (ed.) Rocky Mountain Politics. Albuquerque: The University of New Mexico Press.

EDSALL, PRESTON W. (1962) "North Carolina: people or pine trees." Pp. 98-110 in Malcolm E. Jewell (ed.) The Politics of Reapportionment. New York: Atherton Press.

EPSTEIN, LEON D. (1956) The Wisconsin Farm Vote for Governor, 1948-1954. Madison: The University of Wisconsin Extension Division.

___(1958) Politics in Wisconsin. Madison: The University of Wisconsin Press.

FENTON, JOHN H. (1957) Politics in the Border States. New Orleans: The Hauser Press.

___(1966) Midwest Politics. New York: Holt, Rinehart & Winston.

FLINN, THOMAS A. (1960) "The outline of Ohio politics." Western Political Quarterly (September): 702-721.

FUCHS, LAWRENCE H. (1961) Hawaii Pono. New York: Harcourt, Brace & World.

GARCEAU, OLIVER. (1966) "Research in state politics." Pp. 2-16 in Frank Munger (ed.) American State Politics. New York: Thomas Y. Crowell Co.

___and CORINNE SILVERMAN. (1954) "A pressure group and the pressured: a case report." American Political Science Review 48 (September): 672-691.

GLAZER, NATHAN, and DANIEL PATRICK MOYNIHAN. (1963) Beyond the Melting Pot. Cambridge: The M.I.T. Press.

GOMEZ, RUDOLF. (1969) "Colorado: the colorful state." Pp. 125-151 in Frank H. Jonas (ed.) Politics in the American West. Salt Lake City: University of Utah Press.

HAVARD, WILLIAM C., and LOREN P. BETH (1962) The Politics of Mis-Representation. Baton Rouge: Louisiana State University Press.

HOLMES, JACK E. (1967) Politics in New Mexico. Albuquerque: The University of New Mexico Press.

HUCKSHORN, ROBERT J. (1965) "Decision-making stimuli in the state legislative process." Western Political Quarterly 18 (March): 164-185.

HUNTINGTON, SAMUEL P. (1950) "The election tactics of the Non-Partisan League." The Mississippi Valley Historical Review 36 (March): 613-632.

HUTHMACHER, J. JOSEPH. (1959) Massachusetts People and Politics, 1919-1933. Cambridge: Harvard University Press.

IRWIN, WILLIAM P. (1962) "Colorado: a matter of balance." Pp. 64-80 in Malcolm E. Jewell (ed.) The Politics of Reapportionment. New York: Atherton Press.

JEWELL, MALCOLM E. (1962) "Kentucky: a latent issue." Pp. 111-119 in Malcolm E. Jewell (ed.) The Politics of Reapportionment. New York: Atherton Press.

___and EVERETT W. CUNNINGHAM. (1968) Kentucky Politics. Lexington: University of Kentucky Press.

JONAS, FRANK H. (1940) "Utah: sagebrush democracy." Pp. 11-50 in Thomas C. Donnelly (ed.) Rocky Mountain Politics. Albuquerque: The Univeristy of New Mexico Press.

___(1957) "The 1956 elections in Utah." Pp. 78-87 in Frank H. Jonas (ed.) Western Politics and the 1956 Elections. Research Monograph No. 2, Institute of Government. Salt Lake City: University of Utah.

___(1969) "Utah: the different state." Pp. 327-279 in Frank H. Jonas (ed.) Politics in the American West. Salt Lake City: University of Utah Press.

KEY, V. O., JR. (1949) Southern Politics. New York: Alfred A. Knopf.

___(1956) American State Politics. New York: Alfred A. Knopf.

LAMB, KARL A. (1962) "Michigan legislative apportionment: key to constitutional change." Pp. 267-297 in Malcolm E. Jewell (ed.) The Politics of Reapportionment. New York: Atherton Press.

LARSON, T. A. (1965) History of Wyoming. Lincoln: University of Nebraska Press.

LEE, CALVIN B. T. (1967) One Man, One Vote. New York: Charles Scribner's Sons.

LEVIN, MURRAY B., and GEORGE BLACKWOOD. (1962) The Compleat Politician. Indianapolis: The Bobbs-Merrill Co.

LITT, EDGAR. (1965) The Political Cultures of Massachusetts. Cambridge: The M.I.T. Press.

LOCKARD, DUANE. (1959) New England State Politics. Princeton: Princeton University Press.

___(1964) The New Jersey Governor. Princeton: D. Van Nostrand Co.

McKEAN, DAYTON D. (1967) Pressures on the Legislature of New Jersey. New York: Russell & Russell.

MARTIN, BOYD A. (1969) "Idaho: the sectional state." Pp. 181-200 in Frank H. Jonas (ed.) Politics in the American West. Salt Lake City: University of Utah Press.

MARTIN, CURTIS. (1954) "Colorado." Pp. 60-80 in Paul T. David, Malcolm Moos, and Ralph M. Goldman (eds.) Presidential Nominating Politics in 1952. vol. 5. Baltimore: The Johns Hopkins Press.

MASTERS, NICHOLAS A. and DEIL S. WRIGHT. (1958) "Trends and variations in the two-party vote: the case of Michigan." American Political Science Review 52 (December): 1078-1090.

MITAU, G. THEODORE. (1960) Politics in Minnesota. Minneapolis: University of Minnesota Press.

MOREY, ROY D. (1965) Politics and Legislation: The Office of Governor in Arizona. Tucson: The University of Arizona Press.

MORLAN, ROBERT L. (1955) Political Prairie Fire. Minneapolis: University of Minnesota Press.

MOSCOW, WARREN. (1948) Politics in the Empire State. New York: Alfred A. Knopf.

MOWRY, GEORGE E. (1963) The California Progressives. Chicago: Quadrangle Books.

NEUBERGER, RICHARD L. (1954) Adventures in Politics. New York: Oxford University Press.

NYE, RUSSEL B. (1965) Midwestern Progressive Politics. New York: Harper & Row.

OSTRANDER, GILMAN M. (1966) Nevada: The Great Rotten Borough, 1859-1964. New York: Alfred A. Knopf.

PATTERSON, SAMUEL C. (1968) "The political cultures of the American states." Journal of Politics 30 (May): 187-209.

PAYNE, THOMAS. (1954) "Montana." Pp. 8-25 in Paul T. David, Malcolm Moos, and Ralph M. Goldman (eds.) Presidential Nominating Politics in 1952. vol. 5. Baltimore: The Johns Hopkins Press.

——(1969) "Montana: politics under the copper dome." Pp. 203-230 in Frank H. Jonas (ed.) Politics in the American West. Salt Lake City: University of Utah Press.

PETERSON, HENRY J. (1940) "Wyoming: a cattle kingdom." Pp. 115-149 in Thomas C. Donnelly (ed.) Rocky Mountain Politics. Albuquerque: The University of New Mexico Press.

PETTENGILL, DWYNAL B. (1962) "Maryland: judicial challenge to rural control." Pp. 298-313 in Malcolm E. Jewell (ed.) The Politics of Reapportionment. New York: Atherton Press.

REICHLEY, JAMES. (1964) States in Crisis. Chapel Hill: The University of North Carolina Press.

RICE, ROSS R. (1969) "Arizona: politics in transition." Pp. 41-70 in Frank H. Jonas (ed.) Politics in the American West. Salt Lake City: University of Utah Press.

SARASOHN, STEPHEN B., and VERA H. SARASOHN. (1957) Political Party Patterns in Michigan. Detroit: Wayne State University Press.

SCHLESINGER, JOSEPH A. (1957) How They Became Governor. East Lansing: Michigan State University Government Research Bureau.

——(1965) "The politics of the executive." Pp. 207-237 in Herbert Jacob and Kenneth N. Vines (eds.) Politics in the American States. Boston: Little, Brown & Co.

——(1966) Ambition and Politics. Chicago: Rand McNally.

SHANK, ALAN. (1969) New Jersey Reapportionment Politics. Cranbury: Associated University Presses.

SINDLER, ALLAN P. (1956) Huey Long's Louisiana. Baltimore: The Johns Hopkins Press.

SMITH, C. C., and CHARLES KUNSMAN, JR. (1954) "Nevada." Pp. 100-119 in Paul T. David, Malcolm Moos, and Ralph M. Goldman (eds.) Presidential Nominating Politics in 1952. vol. 5. Baltimore: The Johns Hopkins Press.

SORAUF, FRANK J. (1963) Party and Representation. New York: Atherton Press.

SOUKUP, JAMES R., CLIFTON McCLESKEY, and HARRY HOLLOWAY. (1964) Party and Factional Divison in Texas. Austin: University of Texas Press.

STEINER, GILBERT Y. (1951) Legislation by Collective Bargaining.

Urbanization: University of Illinois Institute of Labor and Industrial Relations.

___and SAMUEL K. GOVE. (1960) Legislative Politics in Illinois. Urbana: University of Illinois Press.

STUMPF, HARRY P., and T. PHILLIP WOLF. (1969) "New Mexico: the political state." Pp. 259-295 in Frank H. Jonas (ed.) Politics in the American West. Salt Lake City: University of Utah Press.

SWARTHOUT, JOHN M., and KENNETH R. GERVAIS. (1969) "Oregon: political experiment station." Pp. 297-325 in Frank H. Jonas (ed.) Politics in the American West. Salt Lake City: University of Utah Press.

WADE, RALPH M. (1969) "Wyoming: the frontier state." Pp. 417-441 in Frank H. Jonas (ed.) Politics in the American West. Salt Lake City: University of Utah Press.

WAHLKE, JOHN C., HEINZ EULAU, WILLIAM BUCHANAN, and LEROY C. FERGUSON. (1962) The Legislative System. New York: John Wiley & Sons.

WALTZ, WALDO E. (1940) "Arizona: a state of new-old frontiers." Pp. 252-291 in Thomas C. Donnelly (ed.) Rocky Mountain Politics. Albuquerque: The New Mexico Press.

WEIBULL, JORGEN. (1965) "The Wisconsin progressives, 1900-1914." Mid-American 47 (July): 191-221.

WIER, JEANNE ELIZABETH. (1940) "The mystery of Nevada." Pp. 88-114 in Thomas C. Donnelly (ed.) Rocky Mountain Politics. Albuquerque: The University of New Mexico Press.

ZEIGLER, HARMON. (1965) "Interest groups in the states." Pp. 101-147 in Herbert Jacob and Kenneth N. Vines (eds.) Politics in the American States. Boston: Little, Brown & Co.

ZELLER, BELLE. (1954) American State Legislatures. New York: Thomas Y. Crowell Co.

INDEX